Sustainable Agriculture Theory and Practices

NIPA® GENX ELECTRONIC RESOURCES & SOLUTIONS P. LTD.
New Delhi-110 034

About the Authors

Sri Bidyadhar Maharana, a Post Graduate in Agronomy from Odisha University of Agriculture & Technology was in government service in Agriculture Department, Odisha from 1965 to 2002 in various capacities. Before his retirement he was a National Professional in a GOI-UNDP Assisted Sub-programme "Strengthening NRM and Sustainable Livelihoods of Women in Agriculture in Tribal Odisha" for five years in Odisha. He was a freelancer Consultant in WB, IFAD, DFID, ADB, UNDP, UNICEF, IWMI, CARE assisted programmes/ projects in Odisha from 1996 to till date. He was a certified Trainer by Department of Personnel and Training, Government of India for Odisha and had trained middle level and high level officials in Odisha on subject matters including Sustainable Agriculture, NRM and Watershed management for about three decades. He was National Monitor of Ministry of RD for three years in connection with monitoring of PMGSY, MGNREGA, IAY, NSAP, IWMP, SBM and NWDPRA programmes in Nagaland, Jharkhand and West Bengal. He is providing consultancy to reputed NGOs/ Organisations like XIMB-CENDERET, CYSD, Agragamee, PRAGATI, NIRMAN, IGSSS, OPSL, RICOR, OTELP etc. till now. He has developed a User Manuals on Micro-Planning, Decentralised GP Planning, GP Minimalistic Plan, Women and Child Friendly GP, RKVY, Nutrition Garden, Land & Water Management Manual for UNDP, UNICEF, SIRD, IGSSS, Government of Odisha. He has written a number of book-lets and reading materials for the farmers and extension workers of Odisha. He was co-author of a book "Integrated Pest Management" in 1990 published by DANIDA Project in Odisha along with Sri G.N.Nayak. He has conducted a number of evaluation studies and Baseline Studies for many organisations like Government of Odisha, IIT, Bhubaneswar, International Water Management Institute (IWMI), CARE India, DFID, NABARD for two decades.

Professor (Dr) Subodha Kumar Sahu, Odisha University of Agriculture & Technology (OUAT), Bhubaneswar was a Professor of OUAT in Soil Science. He was in OUAT service from 1969 to 2002, in research, extension and teaching. After his retirement he was a member of the Senate of agriculture faculty of the University, Viswabharati, Shantiniketan for one year and an Emeritus Scientist of Indian Council of Agriculture Research (ICAR) for one year. He guided 10 PhD students for doctorate degree in Soil Science and Agricultural Chemistry and advised 7 students for PhD. He was also a recognised guide for PhD in Utkal University, Bhubaneswar. He has published 130 articles in various Journals including International Journals. He is author of 12 Books on Agriculture Chemistry and participated in several International Seminars. He is still writing several agricultural books in Odia language for farmers of Odisha. His latest book released in 2021 is on Soil.

Sustainable Agriculture Theory and Practices

Bidyadhar Maharana
S.K. Sahoo

NIPA® GENX ELECTRONIC RESOURCES & SOLUTIONS P. LTD.
New Delhi-110 034

NIPA® GENX ELECTRONIC RESOURCES & SOLUTIONS P. LTD.

101,103, Vikas Surya Plaza, CU Block
L.S.C.Market, Pitam Pura, New Delhi-110 034
Ph : +91 11 27341616, 27341717, 27341718
E-mail: newindiapublishingagency@gmail.com
www: www.nipabooks.com

For customer assistance, please contact
Phone: + 91-11-27 34 17 17
Fax: + 91-11- 27 34 16 16
E-Mail: feedbacks@nipabooks.com

ISBN: 978-81-19235-80-3

Composed and Designed by NIPA.

SIKSHA 'O' ANUSANDHAN

(A Deemed to be University declared u/s 3 of UGC Act, 1956)
Re-Accredited by NAAC with 'A' Grade
Khandagiri Square, Bhubaneswar - 751 030, Odisha, India
Cell: 073810 36054, 094370 40057
Email: chancellor@soa.ac.in,
dpray1949@gmail.com, www.soa.ac.in

Prof. (Dr.) D.P. Ray

FNAAS, FNABS, FHSI, FISVS, FCHI, FftLPUB, FRFRI, FOES
Former Vice-Chancellor, OUAT

Bhubaneswar
7th June, 2022

Foreword

In the initial years of Green Revolution, India's food production grew at an unprecedented scale and the country reached a stage of self-sufficiency in food grain production. High yielding varieties of seeds, chemical fertilisers, assured irrigation and pesticides were key components of this high-input technology. However, the situation has changed over the years with several consequences like groundwater depletion, land degradation, yield stagnation, loss of agri-biodiversity and the long-term impact on farmers' and consumers' health. The problem has been further accentuated by the climate change which directly affects agriculture. Agricultural sustainability has been an issue which has attracted the attention of agri-scientists, policymakers and farmers. With this backdrop the concept of sustainable agriculture has been developed.

Sustainable agriculture involves conscious design and management procedures that works with natural processes to conserve all natural resources, minimize waste and environmental damage, while maintaining or improving agricultural profitability. In practice such systems have aimed to avoid the use of synthetic fertilizers, pesticides, growth regulators, and livestock feed additives. Instead, sustainable agriculture systems rely on crop rotations, crop residues, cover cropping, animal manures, legumes, green manures, off-farm organic wastes, minimum soil tillage, and mineral bearing rocks to maximize soil biological activity, and to maintain soil fertility and productivity. Natural, biological, and cultural controls are used to manage crop pests, weeds and diseases. Sustainable agriculture would provide a long-lasting solution to the climate change which provides a threat to the agricultural production.

The authors, namely Dr S.K.Sahoo and Sri Bidyadhar Maharana have long experience in agriculture with regard to soil science and crop agronomy. Dr Sahoo was a Professor of Soil Science in OUAT and he has more than five decades of experience in teaching, research and field investigation. Sri Bidyadhar Maharana is a post graduate in Agronomy from OUAT and he has more than five decades of experience in extension, research and consultancy in various government, non-government and bilateral organisations like UNDP, DFID, ADB, UNICEF etc.

Both the authors have written this book having 12 chapters which include alternative agriculture, soil-plant-water relationships, integrated pest management, integrated nutrient management, diversified farming system, conservation agriculture, nutrition sensitive agriculture and climate resilient agriculture and practices for sustainable agriculture etc.

I hope, the book will be extremely helpful to the students of agriculture, researchers, agricultural practitioners and those appearing in competitive examinations.

(Dr. D. P. Ray)
Vice-Chancellor

Preface

With the advent of green revolution in 1967, India has substantially increased its agriculture production and productivity in foodgrains which brought a paradigm shift from begging bowl situation to a self-sufficiency stage. But the growth has been more pronounced in irrigated tracts and in large holdings creating an imbalanced growth. Rainfed areas and small holders have been overlooked from such agricultural development. At the same time there have been increased ecological and environmental problems as a result of overuse of fertilisers and pesticides in intensive agriculture. The deteriorating soil and water quality and the rising pollution in agro-ecosystem has been a serious concern. The problem has been further accentuated by the climate change. The growth in agriculture has been stagnant and farmer's income is at declining stage. The smallholder agriculture has not been rewarding and remunerative in many cases for which the farmers are trying to find alternatives. Sustainability of future agriculture has become a matter of concern.

Agricultural sustainability in its broader sense means ecological sustainability. The transformation of agriculture from the situation of 'farming for subsistence' to 'farming for profits' with the backing of technological advancements in improved crop genotypes and animal breeds, fertilizers, pesticides, irrigation and mechanization has resulted in meeting the expected demand of farmers from agriculture. At the same time it has negative impacts of decline in resource base both in quantity and quality. Advances in the science of ecology in the 1970s highlighted the interaction and interdependence of the natural resources of land, water, air, and biodiversity and the life support systems they provided for our ecosystem and their capacity to provide life support, which are not infinite and renewable. Sustainability is the key to preventing or reducing the effect of environmental issues. Ecosystems are dynamic interactions between plants, animals, and microorganisms, and their environment working together as a functional unit.

The goal of sustainable agriculture is to meet society's food and textile needs in the present without compromising the ability of future generations to meet their own needs. The ultimate objective should be the optimum mix of agricultural practices, both old and new, in order to maximize sustainable output within the limits of available resources.

Practitioners of sustainable agriculture seek to integrate three main objectives into their work: a healthy environment, economic profitability, and social and economic equity. Sustainable agriculture is both a philosophy and a system of farming. It has its roots in a set of values that reflects an awareness of both ecological and social realities. It involves design and management procedures that work with natural processes to conserve all resources, minimize waste and environmental damage, while maintaining or improving farm profitability. Sustainable agriculture systems are designed to take maximum advantage of existing soil nutrient and water cycles, energy flows, and soil organisms for food production. Such systems aim to produce food that is nutritious, without being contaminated with products that might harm human health.

In practice such systems have tended to avoid the use of synthetically compounded fertilizers, pesticides, growth regulators, and livestock feed additives. These substances are rejected on the basis of their dependence on non-renewable resources, disruption potential within the environment, and their potential impacts on wildlife, livestock and human health. Instead, sustainable agriculture systems rely on crop rotations crop residues, animal manures, legumes, green manures, off-farm organic wastes, appropriate mechanical cultivation, and mineral bearing rocks to maximize soil biological activity, and to maintain soil fertility and productivity. Natural, biological, and cultural controls are used to manage pests, weeds and diseases.

As in conventional agricultural systems, the success of sustainable approaches is very dependent on the skills and attitudes of the producers. The degree to which different models of such farms are sustainable is very variable, and is dependent on the physical resources of the farmer, and the degree deficiencies in support farm, the talents and commitment of the support available.

There have been many attempts at government and non-government level to promote sustainable agriculture. Various new schemes are implemented in a mission mode in the name of National Mission for Sustainable Agriculture (NMSA) by the Government of India. The present book examines the current practices of sustainable agriculture from production to consumption, and the urgent need to transition to long-term sustainability. It highlights the application of Agro-ecology for developing alternatives to the complex problems of resource depletion, environmental degradation, a narrowing of agro-biodiversity, consolidation and industrialization of the food system, climate change, and the loss of farm land.

The objectives of this book are: (i) to understand the role and challenges of sustainable agriculture and its importance to the sustainable soil managements, (ii) to restore soil health to transforming agriculture for sustainability, and (iii) to understand the principles and practices of alternative agriculture, integrated nutrient and pest management, soil and water management under the backdrop of climate change.

There are twelve chapters in this Book. The **Chapter-1** is introductory which gives the basics of agriculture and food security. **Chapter-2** deals with basics of sustainable agriculture and alternative agriculture that includes organic farming, ecological farming, bio-dynamic farming, natural farming and zero budget natural farming (ZBNF) etc are dealt in **Chapter-3. Chapter-4** describes soil and plant nutrients that are required for plant growth. Water management and dryland technology for sustainable crop production are dealt in Chapter-5. Chapter-6 explains the integrated pest management tactics while **Chapter-7** deals with diversified farming systems and integrated farming systems. Chapter -8 deals with sustainable natural resource management and **Chapter-9** describes the nutrition sensitive food system in the context of sustainability. **Chapter-10** deals with conservation agriculture and **Chapter-11** focuses on climate resilient agriculture for sustainable crop production. **Chapter-12** summarise sustainable agriculture practices. All chapters are well-illustrated with appropriately placed data, tables, figures, and photographs and supported with extensive and most recent references.

We hope the readers will go through the chapters carefully and offer their valuable suggestions which would provide scope for further improvement. References have given at the end of each chapter to enable the reader for further reading.

1st December, 2022
Bhubaneswar

S.K. Sahoo
Bidyadhar Maharana

Contents

1

Introduction

Agriculture depends mainly on natural resources like land situation, soil, water and a favourable climate. Continuous agricultural practices adopted without conservation of such precious resources may lead to unsustainability of growth in agriculture. Conservation of these resources and their sustainable use with appropriate measures would be required for a sustainable agriculture. The government policy for food grain self-sufficiency is not necessarily aims at agriculture sustainability. The growth of agricultural production and productivity cannot be achieved without agricultural sustainability.

1.1 Green revolution in India

Green revolution in India started in 1967 to make the country self-sufficient in food grain production. Mexican dwarf wheat varieties developed by the International Maize and Wheat Improvement Centre (CIMMYT) and High Yielding Varieties of rice evolved in the International Rice Research Institute (IRRI), Philippines were cultivated under irrigated condition with application of fertilisers and pesticides. With the advent of green revolution in 1967, there was a change in India's status of begging-bowl to self-sufficiency in food grains. Green Revolution helped averting famines in India and Pakistan and Norman E. Borlaug, the father of dwarf genes, earned the Nobel Peace Prize in 1970. Green revolution had four basic elements: (i) use of high yielding cultivars, (ii) double cropping in existing farm lands with development of irrigation, (iii) continued expansion of farm areas, and (iv) improved agronomic techniques with use of agro-inputs and mechanisation.

The success of green revolution in India can be judged from the quantum jump in production and productivity of wheat and rice, substantial reduction in food imports, and increased employment in farm sector. However, several drawbacks of green revolution were noticed. The success was confined to irrigated states like Punjab, Haryana, Western UP and Tamil Nadu, whereas many states including Eastern regions remained untouched thereby creating regional imbalance. The plan was implemented only in areas with assured supplies of water and the means to control it and the farmers could use fertilisers, pesticides and farm credit. The green revolution also increased income disparities among the states due to variation in use of such inputs. Use of fertilisers and irrigation without taking care of soil health led to soil degradation and environmental problems. It was also observed that many traditional crops except wheat and rice were not benefitted by the green revolution. Mechanisation was responsible for creating unemployment in some areas. Many farmers who had tried to take on the new technologies became heavily in debt, leading to increase in family stress. The impacts of green revolutions can be summarised as follows:

Ecological and social impacts: Since the indigenous crop varieties were unable to utilise higher dose of fertilisers for better yields, the high yielding varieties (HYVs) were mostly preferred by the farmers which required chemical fertilisers, pesticides and irrigation in excess quantity that led to soil degradation and depletion of water resources. The HYVs with a narrow genetic base have shown yield decline over a period of time and lowered the quality of food grain. Loss of landraces, soil nutrients, leaving toxic residues in food crops and increased farming expenses by the farmers are major socio-economic impacts.

Impact on food grains: Success of wheat, rice and maize has impacted coarse cereals and millets in terms of coverage and production that affected availability of nutritious cereals in rural and urban areas. Availability of minor millets and pulses was decreased while availability of wheat and rice increased. Low consumption of pulses and millets has caused malnutrition.

Impact on nutrition: Millets are more nutritious than rice and wheat with regard to protein, essential minerals and vitamins. Milled rice of HY varieties loses riboflavin, thiamine, niacin, calcium, phosphorus, iron, and zinc during polishing. Consumption of rice, wheat and pulses without nutritious millets, fruits and vegetables in the diet leads to deficiency of micronutrients such as iron, zinc, calcium, vitamin A, folate, and riboflavin as a result of which diseases like anaemia, night blindness and infertility are caused.

Extinction of indigenous crops: Monoculture of cereals such as rice and wheat has replaced the coloured rice, aromatic rice, and medicinal rice varieties, and millets of indigenous varieties which were resistant to adverse situations like drought, salinity, and flood in addition to health benefits. Millets have some medicinal properties also. It is now felt for revival of such indigenous crops for nutrition security.

1.2 Conventional agriculture

Adoption of conventional agriculture or modern/industrial agriculture practices has increased agricultural production and productivity to meet the demand of the growing population and the growth in the sector. With technological interventions and more capital investment the system was able to consume high yielding varieties and hybrids with agro-chemicals for increased production. However, modern agriculture consumed finite resources such as water, plant nutrients and fossil fuels, which caused several ecological problems as outlined below:

- Gradual decline in soil productivity is noticed due to soil and water erosion, soil compaction, loss of soil organic matter and water holding capacity, loss of biological activity and other soil related problems.
- The agricultural practices with adoption of modern technologies have led to soil and water contamination that sediments salts fertilizers (nitrates and phosphorus), pesticides, and manures. Pesticide contamination of ground water and reduced water quality impact agricultural production, drinking water supplies, and fishery production.
- Water scarcity has been observed in many places due to overuse of surface and ground water for irrigation with little concern for the hydrological cycle that maintains stable water availability.
- Loss of biodiversity in almost all the eco-systems has upset natural balance. There has been extinction of many types of flora and fauna.
- Use of excess dose of chemical fertilisers in HYVs has made the crop luxuriant and susceptible to pest attack and upset the natural resistance in crops. This required use of chemical pesticides to protect the crops from pest attack.
- Indiscriminate use of toxic pesticides has led to development of resistance to chemicals in pests and destruction of beneficial insects and pollinators which upset the natural balance in agricultural eco-system. In absence of natural defenders there has been pest resurgence in crop fields leading to pest outbreak and pest endemics. Besides, loss of wetlands and wildlife habitat and reduced genetic diversity due to reliance on genetic uniformity in most crops and livestock breeds are some of environmental concerns.
- Emission of greenhouse gases (GHG) such as carbon dioxide, methane and nitrous oxide etc. has led to climate change with consequence of destruction of tropical forests and other native vegetation for agricultural production. The modern agriculture has been the cause of climate change to a great extent due to emission of GHG.
- Establishment of agro-chemical industries in large number has made the surrounding air and water polluted causing a chain of health hazards to the animal and human population. Indiscriminate use of agro-chemicals by the farmers has encouraged the industry to grow.

1.3 New Challenges for Indian agriculture

Nearly 83% of farmers of the country are small and marginal and the land distribution is skewed. The land-man ratio is declining over the years. Fragmented and scattered holdings (as is the case of most of the marginal and small farms in India) do not allow better utilization of farm resources and technology adoption by the farmers which reduces the productivity. Moreover, this also hinders diversification process which is considered a key in enhancing the income of farmers.

The major challenge for Indian agriculture is to increase crop production and productivity sustainably without any detrimental effect on the soil environment. At the same time profitability is to be sustained to make the farming rewarding. The task has been more difficult to produce enough food for the burgeoning population with shrinking natural resources like land and water. According to Ministry of Agriculture (GoI, 20001a) approximately 53 per cent of geographical area of the country had been considered degraded by early 1980s. As per the report six per cent of the cultivated area was waterlogged, and three per cent was affected by soil alkalinity and acidity. National Agriculture Policy (NAP, 2000) has stressed the importance of management and conservation of resources. The major challenges requiring priority are outlined below:

Raising agricultural productivity and growth: With shrinkage of land and water resources and adverse impact of climate change, raising agricultural growth and productivity has been a major challenge. The agricultural productivity requires sustainability as many areas and many crops have shown yield stagnation and reduction. Appropriate measures like sustainable use of resources, horizontal and vertical diversification to high value crops and improvement in value chains of agricultural produces have become imperative.

Increasing agriculture and non-farm employment: Agriculture is a growth engine for rural development which benefits rural households including poor, landless, agricultural labourers and vulnerable sections. Rural poverty can be reduced by creating more employment in agriculture and non-farm sector. Skill development in agriculture sectors both for on-farm and non-farm employment would be required to reduce rural poverty.

Reforming agricultural research and extension for technology transfer: The investment on R& D and extension is inadequate in the country for ensuring reforms in research & extension for which external funding support has been made by World Bank and other agencies from time to time. Since these services are affected due to inadequacy of funding, personnel and infrastructure, there is a need to enhance R & D investment to at least 2-3% of agricultural GDP.

Ensuring Food & Nutrition security: After showing enhanced growth in 1970s to 1980s on successful implementation of green revolution, the agricultural growth declined in the 1990s and 2000s, which is a major concern. The yields of major crops of India are still lower than other countries like China, Vietnam, Australia, USA and Indonesia. There is a need for increasing competitiveness and diversification in agriculture sector. The Food & Nutrition Security has to be ensured to improve the dubious distinction of lower global hunger index. Diversification to high value crops should be facilitated by providing infrastructure for storage, marketing, transport, processing and export. Allied agriculture sectors like livestock production and aquaculture have high growth potential which are essential for raising farmer's income and ensuring nutrition security.

Natural Resource Management: India with 2.4 per cent of the world's total area and 16 per cent of the world's population has only 4 per cent of the total available fresh water. Agriculture is the biggest consumer of fresh water and has to face increasing competition for water from industry, power generation and domestic use. In India 80-90 per cent of both surface and ground water is used for irrigation with water use efficiency of about 30-35%. On the other hand, about 50% of area under cultivation remains rain-fed. There is gradual declining in ground water level. Integrated Water Resource Management (IWRM) on river basin basis and river-linking projects need to be prioritised. Necessary technologies and measures are required to increase water use efficiency and water productivity. Participatory irrigation and drainage management along with higher investment on operation & maintenance of irrigation systems and cost recovery are to be within the irrigation policy.

Land degradation is caused by both natural causes and human activities and this is a major cause for reduced crop production. Rehabilitation of degraded land, reclamation of problematic soils, control of soil erosion, rain water harvesting & conservation, and watershed development need special attention. Conservation of forest & biodiversity along with afforestation and agro-forestry should be the prioritised interventions for natural resource management for a sustainable agriculture. National Agriculture Policy (NAP, 2000) has stressed the importance of management and conservation of natural resources for sustainable agriculture.

Sustaining the environment and future agricultural productivity: Depletion in ground water level due to over exploitation has been seen in many areas. Creating a water-logging situation in command areas due to over irrigation that develops soil salinity and flooding is a major problem in command areas. In rain-fed areas, soil erosion, land degradation and moisture stress at critical crop growing period make the farming unsustainable. Intensive cultivation with high yielding crop varieties has led to mono-cropping of staple crops like rice and wheat at the expenses of other crops, which has caused loss of agricultural biodiversity. Overgrazing in forest land and uncontrolled deforestation need to be arrested and massive afforestation and compensatory afforestation programme should be taken up.

Climate resilient agriculture: Due to climate change, weather extremities such as droughts, floods, erratic rains have occurred frequently and agriculture has been seriously impacted, particularly in in rain-fed areas. The climate resilient technologies like suitable adaptation and mitigation measures emanating from the research have to be adopted in climate smart villages.

1.4 Agriculture and food security

Food security is defined by the United Nations' Committee on World Food Security as a state when all people, at all times, have physical, social, and economic access to sufficient safe and nutritious food that meets their food preferences and dietary needs for an active and healthy life. There are four dimensions of food security: availability, access, food use& food utilisation and stability

Availability: Food availability can be achieved at national level through domestic production, import, food aids, buffer stock and distribution. This can be achieved at household level through own production, purchase from the market and public distribution system. Food production holds the key to make food available at the national and household level by sustainable agriculture measures and proper utilisation of natural resources.

Access: Access refers to the resources at household level to buy food in proper quantity and quality which speaks about purchasing power of the household. Access could be economic, social and environmental. Economic access depends on the purchasing power and food cost in the market. The price of the food commodities in the market depends on demand and supply. During natural calamities the harvest volume decreases which results in increase in price.

Food use and utilization: Food use and utilisation require availability of drinking water, sanitation, hygiene, education and physical health care. In order to have better food utilisation or absorption, there is a need for healthy physical environment and good sanitation including knowledge about proper preparation, storage, nutrition and health care. Safe drinking water is very important in this context. This dimension of food security requires awareness, education, health and sanitary measures.

Stability: Stability is ensured when food availability or access is achieved at constant level during different periods of the year on long-term basis. Stability also seeks to minimize external risks like natural disaster, price volatility, social conflicts and epidemics. The household should have resilience measures like insurance against crop failure. Protection of environment and sustainable use of natural resources and adoption of climate smart agriculture technologies are required to ensure stability in food security.

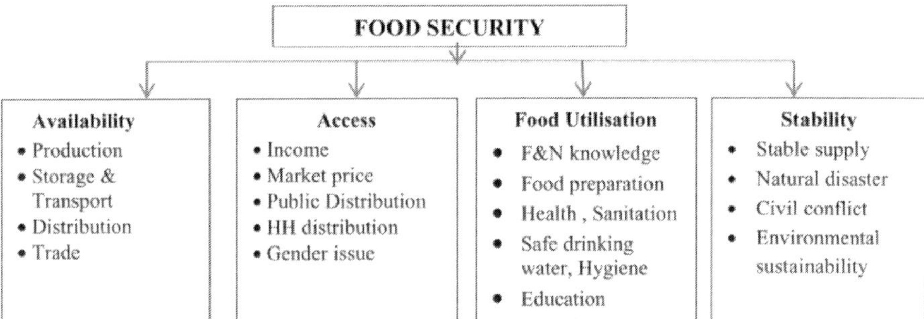

Fig. 1.1: Four dimensions of food security

High population growth requires increase in food production with shrinking resources. By 2050, food production is projected to increase globally by 70% to meet the demand of the projected population. Although 30% of the earth's land is used for agricultural purposes, lack of arable land in many regions has led to the intensification of land with crop potential. Intensive cultivation practices without proper soil health management can reduce the soil productivity and damage soil ecosystem.

As the rate of population increase is especially high in many developing countries, there is need to produce enough food to feed such projected population. Owing to the increasing population pressure over the years, demand for food is naturally expected to increase in coming years. Various studies have projected the demand of food grains under alternative assumptions of income growth, distribution of income and future dynamics of rural and urban populations. Table 1.1 presents projected national demand of major food commodities for the years 2030 and 2050 as estimated in different studies. Substantial increases in the consumption of high-value food commodities like fruits, vegetables, milk, meat, fish and eggs have been projected.

Table 1.1. Projected demands of major food commodities in India (Million tons)

Commodities	2030	2050
Cereals	284	359
Pulses	26.6	46
Edible oil	21.3	39
Vegetable	192	342
Fruits	103	305

Milk	170.4	401
Sugar	39.2	58
Meat	9.2	14
Egg	5.8	10
Fish	11.1	22

Source: Kumar *et al.* (2016) for projected demand in 2030, NCAP Vision 2050 for projected demand in 2050

The present and projected level of aggregate demand for food commodities has been estimated by Ramesh Chand, 2019 which is presented in Table 1.2. As per the estimate, India currently produces about 726 million tonnes of food to meet the food demand of 1.3 billion people. In next 15 years, 40.7 per cent more food will be required to meet domestic demand (Chand, Ramesh, 2019).

Table 1.2. Current production and demand for aggregate food commodities towards 2031-32

Current food production 2015-16	720 million tonnes
Annual growth in demand in next 15 years	2.3 per cent
Total increase in food requirement by 2031-32	40.7 per cent
Quantity of food requirement by 2031-32	1016 metric tonnes

Source: Ramesh Chand, 2019

1.5 Agriculture and eco-system

Agriculture is a fundamental human activity that depends intrinsically on natural processes, including soil fertility, water recycling, pollution, and both nature and agriculture are increasingly suffering the negative impacts of climate change (EEA, 2019). At the same time, unsustainable agriculture also poses a major threat to biodiversity, negatively affects the state of our soil and water, and is an important contributor to climate change (Jia et al., 2019).

The greatest challenges for sustaining the present status and to meet the necessities of future generations are the diminishing carrying capacity of the ecosystem services of our planet in all its ecologies. Almost 40% of crop land has been degraded in less than four decades, and almost 25 billion tonnes of top soil was being lost to erosion every year in the world. Severe depletion of fossil fuel and water shortages in many countries became a cause for worry.

Although intensive agriculture with application of agro-chemicals has resulted in higher productivity, it has made the agriculture unsustainable in the long run due to soil degradation and environmental problems which have impacted human health. Agriculture became the major source of pollution, with nitrate and pesticide residues polluting even ground water adversely affecting the health of the ecosystem. Monoculture with high yielding varieties causes erosion of genetic resources and serious pest attack that require application of pesticides.

The major breakthrough in creating public awareness by Rachel Carson (1962) documented in Silent Spring against the ill effects of pesticides (DDT) on environment and health caused for incorporating environmental concerns of sustainability of modern technologies. There was a growing concern on adverse effects of toxic chemicals which led to the concept of Integrated Pest Management (IPM) as an alternative and rely on ecological principles of pest management.

The most noteworthy contribution to sustainable agriculture came from the system of regenerating/alternate agriculture, which evolved in parallel with industrial agriculture in the United States. New technologies were used for sustainable crop production without any pollution in the soil environment. This was the real birth of a science-based paradigm of sustainable agriculture.

Conservation of resources critical for agricultural productivity includes soil, water and bio-diversity. Long-term sustainability requires maintenance of a healthy soil as a healthy soil promotes healthy crops and livestock. Soil health depends on the quantity of organic matter in the soil and presence of micro-organisms. Organic matter is very important for increasing water holding capacity and microbial activity. It can mitigate the increase of atmospheric CO_2 and therefore climate change.

Recycling of nutrients facilitated by a diversified integrated agriculture and reliance on organic nutrients are important elements of sustainable agriculture. A sustainable agriculture approach should aim to utilize natural resources in such a way that they can regenerate their productive capacity without leaving any adverse effect on the agricultural ecosystem.

References

Agricultural Census (Various rounds), Department of Agriculture and Cooperation, Ministry of Agriculture, New Delhi

Carson, R. 1962. Silent Spring- Pub Houghton Mifflin

Chand, Ramesh (2019).Transforming Agriculture for Challenges of 21st Century: Presidential Address in 102 Annual Conference of Indian Economic Association, 27-29 Dec, 2019

Dalwai, A (2017): Report of Committee on Doubling Farmers' Income Volume I: Ministry of Agriculture & FW

DES, DAC&FW, Ministry of Agriculture & FW, GoI: Agriculture Statistics at a Glance, 2019

EEA technical report No.9/2019, Copenhagen (http://www.eea.europa.eu/publications/eu-emission-inventory-report-lrtap)

FAO (2008): An Introduction to the Basic Concepts of Food Security

Government of India (2000). National Agriculture Policy ,2000, Ministry of Agriculture, New Delhi.

Government of India (2001a), India: Nation Action Programme to Combat Desertification, Volume -I. Ministry of Environment and Forests, New Delhi

Government of India (2002), Tenth Five Year Plan 2002-2007, Volume – II, Planning Commission, New Delhi

Jia, G. et al. (2019): Land–climate interactions. In: Climate Change and Land: an IPCC special report on climate change, desertification, land degradation, sustainable land management, food security, and greenhouse gas fluxes in terrestrial ecosystems

Kapadia (2016): Relevance of Traditional Indian Methods of Water Management in the Present Era India Water Week.

Kumar, P. (2016): "Demand vs. Supply of Food in India-Futuristic Projection." Proceedings of the Indian National Science Academy 82.5: 1579-1586

14.NCF (2007): Report of National Commission on Farmers, Department of Agriculture & Cooperation, Ministry of Agriculture, Government of India

NRSA (2002), Wastelands Atlas of India, Department of Space, Hyderabad, India

2

The Basics of Sustainable Agriculture

2.1 Effects of agro-chemicals used in modern agriculture

Modern farming system was aimed to maximise crop production by use of high yielding crop varieties or hybrids and application of agro-chemicals like fertilisers and pesticides along with irrigation and better crop production technologies. But without consideration of its impact on the environment and economics of the farmers, the monoculture of rice and wheat with agro-chemicals caused land degradation, soil and water contamination, gene erosion and environmental pollution. The agricultural ecosystem services like replenishment of soil productivity and nutrient cycles were greatly disturbed and farming was not sustainable.

Modern agriculture has caused some adverse impacts in the environment and affected both human and animal health. The high yielding varieties made the crop luxuriant after application of fertilisers and micro-nutrients, but such crop plants became susceptible to the attack of insect pests and disease pathogens. In order to control such pests farmers used synthetic pesticides. Such agro-chemicals, though required for yield maximisation to meet the demand of the growing population, disturbed the naturally existing balance between the biotic and abiotic factors in the agro-ecosystem and acted as a potential source of environmental pollution and food contamination. The impact has been noticed as loss of soil fertility, contamination of agro-products, pollution of water bodies and damage to animal and human health.

i) **Fertilisers:** Usually farmers apply more of nitrogen than phosphates and potassium to their crops. Nitrogenous fertilisers applied to soil are adsorbed to the soil particles, taken up by plant roots, utilised by the soil microbes, and leached to the deeper zone of soil or mixed with ground/surface water. Phosphate fertilisers are immediately utilised in limited quantity by plants, while major quantities are fixed in the soil to be available afterwards. Limited quantity of applied phosphate may also be lost to the groundwater. Potassium applied to the soil is directly or indirectly taken up by the crop plants in part, while the remaining part enters into the surface or ground water unless utilised by the microbes.

Only a part of chemical fertilisers is actually utilised by the crop and major part is subjected to various processes of chemical or microbial reactions. The released nutrients by such reactions may be lost through erosion, runoff water or ground water. However, these reactions are influenced by availability of soil moisture, soil pH, organic matter content, microbial population, temperature and uptake by plants. Adverse balance of these factors may increase the loss of nutrients from the soil.

ii) **Pesticides**: Farmers use pesticides and fungicides to protect their crop from attack of insect pests, mites and diseases. Herbicides are used to control weeds in the crop field. The pesticides are either sprayed or dusted on the foliage of crop and the granular insecticides are directly applied to the soil. A part of the applied pesticides/ fungicides/ herbicides reaches the target pests and major part reaches the non-target components of the eco-system such as soil, water, air , soil fauna and other beneficial organisms. An estimated 0.1 to 1% of applied pesticides reaches the target pests, while the remaining part reaches the non-target components and acts as a potential source of pollution of water and soil or take part in metabolic activities of microbial population and soil fauna.

iii) **Effects of agro-chemicals on soil**: Soil is a living dynamic medium for plant growth with integration of physical, chemical and biological spheres. Innumerable soil organisms like bacteria, fungi, actinomycetes, protozoa, algae, annelids, arthropods and mammalians live in harmony with the physical and chemical spheres. Microbial reactions are influenced by physical and chemical properties of soil. The biological cycles such as nitrogen cycle, carbon cycle and hydrological cycles etc. favour adaptation of these microorganisms in the soil. In the soil eco-system there is stability due to food chain and food web of these microorganisms that cause a delicate balance between plants and animals living above the soil. Before application of fertilisers and pesticides the organic matter content of the soil is high due to application of organic matters and recycling of crop residues. But application of agro-chemicals leads to crop susceptibility to pest attack that required chemical pest control. These chemicals have direct and indirect effect on soil health. There is change in soil pH, organic matter content, soil structure & texture, water holding capacity and nutrient availability. Addition of fertiliser for increased crop production may also result in complexities in availability, reaction, transformation and fixation of nutrients. The problem has been further accentuated by alteration of soil properties, climatic factors and changing biological cycles. Fertilisers and pesticides have disturbed the natural processes in soil eco-system and affected the normal life process of soil flora and fauna.

Application of agro-chemicals has adverse effect on ecology of aquifers. The residues of these chemicals have leached into the water bodies including streams, rivers, seas, ponds, wells, estuaries, lagoons, reservoirs around agricultural crop zones. Both eutrophication and contamination of water bodies have been reported. Ground water has also been polluted by application of fertilisers and pesticides to crops. The toxic effects of agro-chemicals have affected the human and animal health directly or indirectly on long-term basis. Finally the food we eat is also contaminated by these agro-chemicals. In order to mitigate the above effects of modern agriculture, the concept of sustainable agriculture has been conceived.

2.2 What is sustainable agriculture?

The word "sustain," comes from the Latin sustinere (sus-, from below and tenere, to hold) that means to keep in existence or maintain and implies long-term support or permanence. In case of agriculture sustainable agriculture is a farming system that meets the demand of the present generation and preserves the resources for the future generation without any environmental problems. Such systems must be resource-conserving, socially supportive, commercially competitive, and environmentally sound (USDA).

Sustainable agriculture can be defined in many ways, but ultimately it seeks to sustain farmers, resources and communities by promoting farming practices and methods that are profitable, environmentally sound and good for communities. It can be defined as a system that can indefinitely sustain itself without degrading the land, the environment or the people. The FAO defines the sustainable agricultural as "the management and conservation of the natural resource base, and the orientation of technological and institutional change in such a manner as to ensure the attainment and continued satisfaction of human needs for present and future generations". Sustainable agriculture enhances the true values of producers and their products. In short, sustainable agriculture is economically viable, socially supportive and ecologically sound.

The definition by G.K. Douglas (1984) formulated little earlier captures the essence of sustainable agriculture in its totality. "Sustainability must be regarded as long-term food sufficiency which requires that agricultural systems be more ecologically based and do not destroy their natural base. Sustainability as stewardship means that agricultural systems are based on a conscious ethics regarding humankind's relationship to future generations and to other species in nature. Sustainability as a community means that agricultural systems are equitable. Agricultural systems cannot be sustained if there is gross misdistribution of land, wealth and power. Social tensions and upheavals in many parts of the world are often manipulations of such inequalities".

Another definition by Mac Rae et al. (1990) argues that, "Sustainable agriculture is both a philosophy and a system of farming. It is rooted in a set of values that reflects an awareness of both ecological and social realities and a commitment to respond appropriately to that awareness. It emphasizes design and management procedures that work with natural processes to conserve all resources and minimize waste and environmental damage while maintaining and improving farm productivity".

Difference between sustainable agriculture and modern agriculture is outlined in Table 2.1.

Table 2.1: Difference between sustainable and modern agriculture

Particulars	Sustainable agriculture	Modern agriculture
Farming system	Diversified or integrated farming system	Monoculture or intensive farming with commodity crops
Plant nutrients	Organic manures, green manures, bio-fertilisers, crop residues and crop rotation	Chemical fertilisers and micro-nutrients
Plant protection	Cultural, varietal, physical and biological methods with crop rotation	Synthetic pesticides
Ecology	Stable ecology	Fragile ecology
Inputs	High diversity, renewable and biodegradable	High productivity, low diversity chemicals and non-renewable
Use of resources	The rate of extraction does not exceed the rate of regeneration	The rate of extraction exceeds the rate of regeneration
Quality of food materials	Safe and quality food materials	Chemical residues are available on food products
Goal	Efficiency, uniformity, maximisation	Environmental health, economic profitability and equity

2.3 Agricultural sustainability

In a simpler sense sustainability means maintainability. In agriculture, sustainability embodies widely varying components, cropping systems and crops grown on different climates with different targets. The sustainability of soil, net profit, eco-system and the farming system must find a place in agricultural sustainability. A crop may produce maintain its productivity over years in an agro-ecosystem, but the whole system cannot be sustainable if it does not achieve the goal of economic sustainability. This may be due to reduced market price or want of proper storage facilities. Therefore economical sustainability is more important than the yield stability.

Intensive use of fertilisers in irrigation command areas may produce sustainable yield in food crops such as rice and wheat, but it may not be sustainable in the long run due to soil degradation. This has happened today in many irrigated areas of the country where either the yield has been stagnated or declined.

2.3.1 Sustainable development and sustainable agriculture

"Sustainable development can be defined as development that meets the needs of the present without compromising the ability of future generation to meet their own needs" (UNESCO, 2015). International organizations such as Food and Agriculture Organization (FAO), the World Bank, and the Consultative Group on International Agricultural Research (CGIAR) made sustainability the basic objective to be pursued in all their future programmes in agriculture. There are some vital reasons for moving away from conventional agriculture and ensure sustainable agriculture as indicated below.

Resource use: Majority of the farmers are small holders with limited resources which are not conducive for intensive agriculture. It is well known that the modern crop varieties require essential nutrients and timely pest control with agro-chemicals. It is also true in case of high producing livestock breeds that require feeding and medical care. In spite of government subsidy, small holders fail to comply with the requirement of such modern crop varieties and livestock breeds. On the other hand, sustainable agriculture with traditional varieties and livestock breeds increases resource use efficiency without dependence on external inputs at higher cost.

Energy use: The farm produce in conventional agriculture require comparatively more energy which is non-renewable. Agricultural machinery used in mechanisation requires fuel received from fossil-based sources. External inputs require transportation from distant places for which non-renewable fuel is used. Fertiliser production accounts for a major share of total energy consumption in modern agriculture. On the other hand, sustainable agriculture reduces dependence on external inputs and limits the energy use.

Environmental safety: In modern and large scale agriculture the precious agro-ecosystem and forest eco-system have been disturbed. The adverse effects of such agriculture have been discernible in contamination of pesticide in food and drinking water, air pollution and emission of greenhouse gases to the atmosphere, occurrence of natural calamities, loss of pollinators and other beneficial organisms, soil erosion, desertification and degradation. All these environmental effects are not seen in sustainable agriculture.

Climate resilience: In absence of diversity, conventional agriculture relies on monoculture of commodity crops or limited breeds of livestock that leads economic production, but lacks in climate resilience. Any weather aberration may cause serious crop/animal loss due to climate change. In such a situation sustainable agriculture is the best alternative. With conservation of biodiversity and maintenance of soil health there is a natural resistance in crops and livestock against drought and flood. Genetic and species diversities increase climate resilience. It is widely seen that carbon sequestration is an advantage of sustainable agriculture.

Food security: Commodity crops grown in conventional agriculture are resource-intensive. Such crops are often used in diversified products as a result of which they are not available fully for human consumption or animal feed. Therefore, conventional agriculture does not allow the community to control their food system.

Sustainable agriculture thus operates within the bounds of physical and biological resources on the one hand and socio-economic viability and quality on the other.

2.3.2 Sustainability in agriculture

The agro-ecosystem and farming system include the complex interactions among their respective components and sustainability is the net result of these interactions. Intensive use of fertiliser is practiced in modern agriculture to maintain the crop yield sustainably over years. But it has detrimental effect on soil due to failure to recycle the organic residues and apply organic matters that has led to unsustainability in many ways. Some of such manifestations are reduced crop yield even after use of recommended fertiliser, artificially maintained yield with excessive and indiscriminate use of fertilisers, increased pest and disease attack despite use of pesticides and increased erodibility of soil due to many factors. Yield plateau is reached in many crops, compelling more investments for which the profitability is also reduced.

Sustainability in agriculture has been a complex phenomenon, where sustainability of yield, profit, soil and whole agro-ecosystem are to be considered concurrently. The least damage to the nature in a given situation is the essential feature of achieving sustainability. As long as natural balance is undisturbed, the sustainability is not at risk. The balance between nutrient uptake and supply, water intake and use in soil, predators and pests, organic sources and microbial population in soil, animal health and crop growth, cattle population and fodder supply are examples of such balances to be interfered in nature to ensure sustainability (Fig. 2.1).

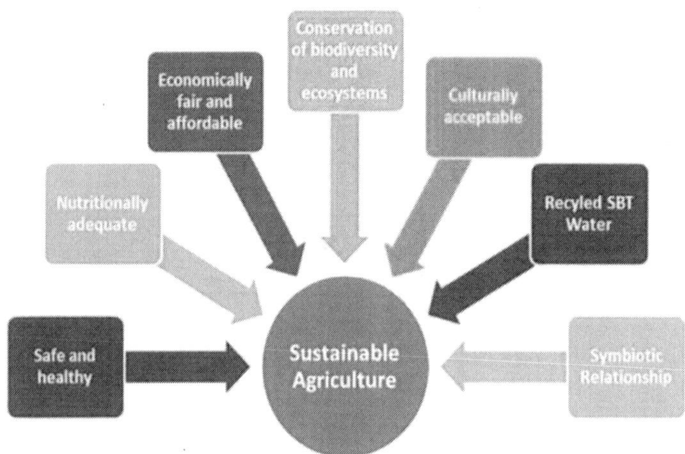

Fig. 2.1. Characteristics of sustainable agriculture

2.3.3 Rationale for sustainability in agriculture

It is now realised that damage to agro-ecosystem is causing irreversible detrimental effects to agricultural production system. Despite increase in crop yield of many crops, the stability in production systems has escaped from modern agriculture. Frequent changes in the cropping system, reduced profitability and unprecedented occurrence of pests are examples of such unsustainability. All the components of agro-ecosystem, such as soil, the environment having the flora and fauna of an agro-ecosystem must live in harmony with each other to achieve the long term sustainability in yield and economic profit. Unless this balance is maintained, the system cannot achieve sustainability as the natural method of supplying nutrients in the living medium would collapse by the imbalance created by the use of modern agro-chemicals. These agro-chemicals have also resulted into number of ecological hazards and health hazards to both human beings and other animals. These ecological hazards such as soil and water pollution, food poisoning, eutrophication of water bodies, and health hazards to both human beings and animals need to be halted. This is only possible through sustainable agriculture that needs:

- To supply of plant nutrients within natural capacity of soil to produce, store and supply;
- To control crop pests by ensuring a natural balance between pests and natural defenders with least human interference;
- To adopt crop sequences and agricultural practices with mutual support and dependence between crops, nutrient demand and controlling carryover of pests from previous to succeeding crop;

- To build the organic status of the soil to produce and supply nutrients to crops and micro-flora status of the soil; and
- To design crop plans for economic and yield sustainability.

2.4 Goal and objectives of sustainable agriculture

Sustainable agriculture is based on diversity of plants and animals in which one component compliments the other. The crops are capable of resisting pest attack and withstanding weather fluctuations. Soil is treated as a renewable resource to produce healthy food without harming the environment. The goal of sustainable agriculture is to meet the present demand of food and fibre without any environmental problems to the meet the demand in future. The main objectives of sustainable agriculture are: (i) a healthy environment, (ii) economic profitability, and (iii) social and economic equity. The specific objectives are to;

- Make optimum use of resources with minimum use of non-renewable resources;
- Enhance and maintain the environmental quality and natural resources;
- Sustain the economic viability of farming;
- Produce enough high quality and safe food; and
- Enhance the wellbeing of farmers and society.

Sustainable farming has fuelled many farming alternatives like organic farming which is a sub-set of sustainable farming. Sustainable agriculture works with nature. It is both a philosophy and a set of sound farming practices.

2.5 Components of sustainable agriculture

If a food production system is not ecologically sustainable, it cannot continue in the long run and cannot be productive and profitable. In order to be profitable and productive, the system should be economically sustainable. Social acceptability, harmony and equality are critical to the sustainability of agriculture or any other human development system (Fig. 2.2).

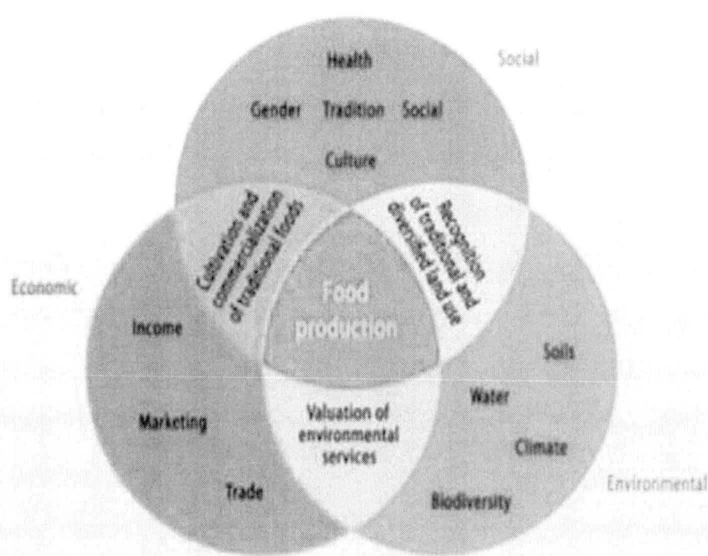

Fig. 2.2. A generalized diagram showing interrelationships between basic components of sustainability (IFOAM)

Productivity: Improvement in crop yield is inevitable under the situations of shrinking resources for agriculture. The double challenges of decreasing per capita arable land and limits to intensification by use of costly external inputs have necessitated a search for newer, alternative approach for increasing both production and productivity within reasonable adverse ecological and socio-economic impacts. This is a concern in terms of the productivity component of sustainable agriculture.

Economic viability: The subsistence agriculture has been shifted to profitable agriculture and the products have shifted from "local to global" to meet the growing demands of people & market and increase profitability. Economic viability is mostly linked to productivity, but it goes with the dynamics and specific need of various income groups of population. When country needs production the farmers want profit. Therefore economic viability for the farmer is essential if agriculture is to be sustainable.

Ecological viability: Modern farming techniques with dependence on monoculture, high yielding varieties/hybrids, fertilisers, agro-chemicals, irrigation, mechanisation etc. have increased the productivity and profitability, but caused some negative impacts on soil health and agro-ecosystem. This has resulted in increased greenhouse gas (GHG) emission, loss of biodiversity, nutrient imbalance, pest resistance & resurgence and climate change. Required infrastructures like warehouse, cold stores, irrigation projects, roads& transport have also modified the ecological balance at regional/local level. In such a situation, conservation agriculture to sustain biological productivity and ecosystem services has been imperative to sustainable agriculture. Conservation of natural resources, reduced use of non-renewable energy & agro-chemicals and maintaining the ecological safety and quality are required for ecological viability of the production system.

Social acceptability: Farming practices and adoption of technologies are required to be acceptable to social norms of the community, while it should have relevance to the macroeconomic policies of the governments and countries. It also encompasses several parameters such as poverty, education, social capital, justice and equity, value systems, food security and livelihood opportunities as per government policy. The technologies also differ as per the resource base of the farmer, who may be predominantly a subsistent or commercial. Thus social justice and equity are the fourth component of sustainability.

Sustainable agriculture seeks permanence to agricultural production systems to ensure food security for all in an ecologically sound, economically viable, and socially responsible manner. In the current globalized and liberalized trade environment, changes that are favourable or adverse can occur over short periods to affect local and international prices and demand for agricultural commodities. Agricultural systems have to adjust to these changes to be sustainable. The alternate approaches like organic agriculture, regenerative agriculture, alternate agriculture, low-input traditional agriculture, biodynamic agriculture, and ecological agriculture have all been promoted as sub-sets of sustainable agriculture. However, sustainability is not inherent in any particular set of practices or farming system, except that some forms of agriculture have more enabling factors for ecological sustainability than others (Raman, 2006).

2.6 Indicators of sustainability

As sustainability is context-specific, common indicators may not be sufficient to measure the agricultural sustainability. RIEDC (1997) indicated general broad measurable components under each level of components of sustainability (Table 2.2). Hoang (2013) had analysed productive performance of crop production systems in an integrated analytical framework considering economic, institutional, physical, social and technological factors and indicated that in a dynamic analysis to make efficiency framework to be forward looking, climate change innovations in crop science to be incorporated.

Table 2.2. Sustainable indicators (Adopted from RIEDC, 1997)

Level	**Sustainability indicators**
Cropping system/Farming system	Non-negative trends in : 1. Farm productivity 2. Net farm income 3. Total factor productivity 4. Nutrient balance 5. Soil quality 6. Residues in soil, plant and products 7. Farm water use efficiency 8. Farmer skill and education 9. Debt service ratio 10. Health 11. Time spent on other social and cultural activities
Agro-ecosystem (watersheds and agro-economic zones)	Non-negative trends in : 1. Regional production 2. Regional income 3. Regional total factor productivity 4. Regional nutrient balance 5. Income distribution 6. Species diversity 7. Soil loss 8. Ground water quality 9. Surface water quality 10. Regional social and economic development indicators
Global, National and Regional System	Indefinitely meet the demands at acceptable social, economic and environmental costs

2.7 Scope of sustainability

An eco-system based approach is required to widen the scope of sustainability in agriculture. It includes good soil health, safe environment, balanced system in flora and fauna in agro-ecosystem, economical profitability of production system besides the sustained agronomic productivity. There should be four types of sustainability such as (i) sustainability of crop returns, (ii) sustainability of net returns, (iii) sustainability of soil fertility, and (iv) sustainability of eco-systems.

Organic farming can achieve greater sustainability than modern chemical farming. The role of organic farming is not confined to supply organic nutrients in place of chemical fertilisers only. The primary role of organic farming extends to improve soil health. It aims to improve the crop yields through the sustainability of ecosystem and soil fertility and makes use of natural processes for pest control as well as decomposition, nitrogen fixation, structural improvement of soil etc. for supply of plant nutrients.

2.8 Strategies for agricultural sustainability

Ensuring ecological sustainability and socio-economic equity are two broad strategies for agricultural sustainability. The specific strategies for sustainable agriculture can be grouped as (i) issues related to farming and natural resource (ii) crop and animal production practices, and (iii) economic, social and political context.

Farming and natural resources management: The basic natural resources of soil and water in a given ecological region are primary sources of production. The rate of exploitation of these resources should not exceed their rate of regeneration. Similarly the wastes should not be generated beyond the assimilation capacity of the environment.

Soil: Soil is store house of plant nutrients and water and acts as an integrator of the environment. It sustains biological activity, diversity, productivity and regulates the nutrient supply and other biological functions for plant growth and development. Soil acts as a major source and sink for global gases and emits greenhouse gas (GHG) to the atmosphere. Soil erosion caused by water and wind is one of the important reasons for soil degradation. Adoption of no tillage could benefit soil conservation in soils especially under initial high fertility conditions (Fabrizzi *et al.*, 2005) without any adverse impact on productivity. Fukuoka (1978) has suggested "Do-nothing Farming" in his book "One straw revolution". Howard (1943) has suggested that "The maintenance of the fertility of the soil is the first condition for any permanent system of Agriculture". Soil management for sustainable production should focus on preventing soil degradation and soil health management. Conventional agriculture disturbs the natural equilibrium among the biophysical components of the soil system than with traditional systems. This can lead to a temporary decline in soil productivity and ecological functions after one or more crop cycles. Soil resilience can be looked upon as the buffering capacity of the soil against physical, mechanical, chemical or biological impacts. Physical, chemical, and biological properties of the soil and soil processes are at the heart of land management for sustainable agriculture.

Water: Agriculture uses more than 80% of freshwater and it is the most critical and limiting natural resource for sustainable agriculture. Overdraft of groundwater for intensive cropping or saving the crop from drought and consequent intrusion of salt water or permanent collapse of aquifers in many regions leading to degradation of potential lands and a serious threat to food security and the environment. Global warming causes erratic rain, drought, and floods accentuate the problem further. Curtailing the share of water that goes to agriculture and ensuring its conservation and sustainable use are essential for future food security, economic and human development and social harmony.

Energy: Modern agriculture is heavily dependent on non-renewable energy sources, especially petroleum products. Sustainable agricultural systems should reduce reliance on fossil-based non-renewable energy sources and explore and exploit many opportunities of on-farm renewable sources of energy generation and use that can also mitigate climate change, as biomass is a carbon-neutral resource.

Crop production practices: Appropriate crop management practices based on the principles of site specific management of soil, nutrient, crop, pest, etc. for responsible farming as per the capability of the resource base increase the value and capacity of resource. Reducing monoculture and adopting diversification can minimise the risk in farming. Crop diversification is helpful to increase crop output under different conditions. Both horizontal and vertical diversification would be required to increase cropping intensity, productivity and income from farming. Diversified farms are usually more economically and ecologically resilient. Cover cropping, crop rotation, mixed and inter cropping and agro-forestry can mitigate the climate risks and ensure sustainability in crop production. Traditional agro-systems balance productivity and sustainability by maintaining high species, and genetic diversity and natural pest control. Diversity of species and genes increases the ability of ecological communities to resist weather aberrations and environmental stress.

Conservation Agriculture: Conservation agriculture relies on managing agro-ecosystem for improved and sustained productivity, increased profits and food security while preserving the resource base and environment. Minimum tillage, permanent soil cover and diversification of species are three principles of conservation agriculture. It enhances biodiversity and natural biological processes above and below the ground surface. Complemented by other known good agronomic practices, including the use of quality seeds, and integrated pest, nutrient, weed and water management, etc., conservation agriculture serves as a base for sustainable agricultural production. It believes that a 'healthy' soil is a key component of sustainability. In sustainable systems, the soil is viewed as a fragile and living medium that must be protected and nurtured to ensure its long-term productivity and stability.

Efficient use of inputs

For sustainable crop production, it is extremely important to maintain adequate levels of such inputs as mineral fertilizers, organic matter and water. Improper management of these inputs has been the case of land degradation and environmental pollution in both low and high-external-input intensive agriculture. The amount of external inputs needed is guided by the efficiency of their use. High input agriculture has seen indiscriminate over use of nitrogen and phosphate fertilizers, which has caused damage to the production resources as well as the environment. The excessive use of fertilizers often leads to wastage and runoff, causing pollution of surface and ground water and leading to nitrate accumulation. Likewise, overuse of water in irrigated agriculture also leads to many problems causing unsustainable production. The best prescription to prevent inefficient use of external inputs is to follow the golden rule of need-based supply of such inputs as tillage, water and fertilizers and to check their wastage. There are proven methods for increasing the use efficiency of different inputs like nutrients and water which must be made use of to attain sustainable crop production.

Animal Production Practices: Integration of crop husbandry with animal husbandry has proved to be sustainable. Integrated farming system (IFS) is designed in such a manner that the output of a system is used as input in the other system which provides scope for organic recycling. In spite of growing specialization of livestock and crop producers, many of the principles of crop production systems apply to both groups. Confined livestock production is increasingly a source of surface and ground water pollutants, particularly where there are a large number of animals per unit area. Expensive waste management facilities are now a necessary cost of confined production systems. In IFS such waste management is not a problem.

References

Douglas, G.K. 1984. Agricultural Sustainability in a Changing World Order. Boulder, CO: Westview Press.

Fabrizzi, K.P., Garcia, F.O., Costa, J.L. and Picore, L.I. 2005. Soil water dynamics, physical properties and corn and wheat responses to minimum and no tillage systems in the southern Pampas of Argentina. Soil and Tillage Research, 81: 57-69.

Fukuoka, M (1978), The One-Straw Revolution: An Introduction to Natural Farming, RODALE PRESS : EMMAUS : 1978.

Haong, Viet-Ngu, 2013. Analysis of productive performance of crop production systems: An integrated analytical framework. Agricultural Systems. 116: 16-24.

Haward, A(1943): An Agricultural Testament: Oxford University Press.

Hegde, D.M & Sudhakar Babu, S.N : Sustainable Agriculture.

Hegde, D.M. and Sudhakara **Babu**, S.N. 2016. **Sustainable Agriculture**. In Modern Concepts of Agronomy (Ed. D.S. Rana, P.K.Ghosh, Y.S. Shivay, Gurubachan Singh).

Mac Rae, R.J., Hill, S.B., Hennings, J. and Bentley, A.J. 1990. Policies, programs and regulations to support the transition to sustainable agriculture in Canada. American Journal of Alternative Agriculture, 5(2): 76.

Raman, S. 2006. Agricultural Sustainability: Principles, Processes and Prospects. Food Products Press, New York, USA, pp. 474.

RIRDC (1997). Sustainability indicators for agriculture, Rural Industries Research and Development Corporation, Australia, 1997, 54 pp.

UC Sustainable Agriculture Research and Education Program. 2021. "What is Sustainable Agriculture?" UC Agriculture and Natural Resources. <https://sarep.ucdavis.edu/sustainable-ag.

3

Alternative Agriculture for Sustainability

3.1 Emergence of alternate farming systems

Alternative agriculture is a production system that follows the concept of agro-ecology and does not use conventional methods. Natural farming, Bio-dynamic farming, Organic farming, Permaculture, *Homa* farming and many others are considered as alternative agriculture. In spite of some differences, these practices aim at preserving the environment and minimise the use or avoid synthetic agro-chemicals. They rely on natural cycles, by using crop rotations, cover crops or no-tillage, etc. The alternative agriculture practices are designed to reduce or eliminate chemical fertilisers and pesticides that are key components of conventional agriculture. The common features are crop rotation with inclusion of legumes and recycling of organic wastes for supply of plant nutrients. In organic systems "natural" sources of inputs are used with discretion (Harwood, 1984).Alternative agriculture, as the term used here, designates farming practices that avoid or at least minimise the use of non-renewable production inputs such as synthetically compounded fertilisers, pesticides and herbicides (Anderson, 1985).

Probably bio-dynamic agriculture by Rudolf Steiner (1861-1925) was the first comprehensive alternate farming system. Sir Albert Howard (1873-1947), a British Botanist, worked as an agricultural advisor in India from 1905 to 1924 wrote a book " An Agricultural Testament" (Howard, 1940) and put forth the " Law of Returns" which called for recycling of all organic waste materials including sewage and sludge back to farm land. He focused on formation of humus in soil as a stable organic product. In fact the present form of organic farming was developed from the philosophical views of Rudolf Steiner, Sir Albert Howard and Lady Eve Balfour. Some of the other variants of alternate farming such as Rishi Krishi, *Homa* farming and Homoeopathic farming are also in vogue (Chhonkar and Dwibedi, 2004). A recent addition of Zero Budget Natural farming (ZBNF) has been promoted by Padmshri Subash Palekar.

3.2 Goals and practices of alternative agriculture

Alternative farming practices include many practices such as Integrated Pest Management (IPM); low intensity animal production system; crop rotation with legumes for reduction of pests, improving crop health, decreasing soil erosion, fix atmospheric nitrogen in the soil : and minimum tillage for reduction of soil erosion and help control of weeds. The broad goals of alternative agricultural practices are to:

- Encourage natural processes such as nutrient cycles, biological nitrogen fixation and pest management through pest-predator relationship ;
- Reduce use of off-farm inputs for safety of the environment and health of farmers and consumers:
- Use of the biological and genetic potential of plants and animal species:
- Keep the cropping pattern in harmony with the limitations of agricultural land to ensure sustainability of production system; and
- Ensure profitable and efficient production system with emphasis on conservation of soil, water, non-renewable energy and biological resources.

Practices

Various practices are adopted in different alternative farming systems keeping the broad goals in view. The important practices are:

- Adoption of crop rotation to reduce soil erosion, pest attack, weed infestation and to increase availability of biological nitrogen;
- Integrated Pest Management (IPM) to reduce use of toxic pesticides by crop rotation, cultural and biological methods , scouting, weather monitoring and use of resistant/tolerant crop varieties;
- Adoption of management system for controlling weeds and improving the soil and plant health and ability of crop plants to resist the insect pest and diseases;
- Use of appropriate measures for soil and water conservation:
- Encourage animal production system emphasizing on health of animals by preventing diseases so as to reduce the use of medicines and antibiotics: and
- Genetic improvement of crops for efficient use of nutrients and resist crop diseases and insect pests.

3.3 Alternative farming systems

Considering the adverse effects of industrial/conventional agriculture, there has been a movement for alternate farming systems. There is an urgent need for identifying potential alternative farming strategies to achieve long term sustainability and food security as indicated by several leading workers in the field. The alternative farming techniques that were field tested and perfected over several generations in the past portrayed the following advantages over chemical farming:

i) Eco-friendly by protecting and reviving life support systems and ecosystem services,
ii) Higher cost benefit ratio, benefiting the farmers as well as the consumers,
iii) Control and reduction of bio-accumulation and bio-magnification,
iv) Reduction in air, water and soil pollution caused by various pesticides and other chemicals,
v) Control of health hazards in humans and livestock, and
vi) Conservation and sustainable use of on-farm biodiversity, including traditional cultivated germplasm and natural resources in agro-ecosystems.

Alternative farming techniques like organic farming, permaculture, polyculture, bio-dynamic farming, eco-farming, no-till farming, integrated farming systems, natural farming and zero budget natural farming are major alternative farming systems that are found to be sustainable.

3.3.1 Organic farming

Organic farming is an agricultural system that does not use any fertilisers and pesticides. It is a production system that avoids use of synthetic pesticides, herbicides, fertilizers, hormones, etc. and relies on techniques like crop rotation, recycling of organic wastes, farm manure, bone meal, rock additives and crop residues for plant protection and nutrient utilization. Thus, in other words, organic farming is relying on the natural process of yield benefits so as to maintain a healthy soil, eat healthy food and grow healthy humans. Organic farming uses organic manures mostly from animal manures, decomposed organic waste and biological nitrogen and ecologically based pest controls methods. Compared with conventional agriculture, organic farming is found to be environmentally safe and does not pollute the soil environment, water and does not create any health hazard for human beings and animals. The lower yield than the conventional agriculture is compensated by higher food costs and environmental cost.

The concepts of organic agriculture were developed in the early 1900s by Sir Albert Howard, F.H. King, Rudolf Steiner, and others who believed that the use of animal manures (often made into compost), cover crops, crop rotation, and biologically based pest controls resulted in a better farming system. The demand for organic food was stimulated in the 1960s by the publication of *Silent Spring*, by Rachel Carson, which documented the extent of environmental damage caused by insecticides.

International Federation of Organic Agriculture Movements (IFOAM), an international organization established in 1972 for organic farming organizations defines the goal of organic farming as:

"Organic agriculture is a production system that sustains the health of soils, ecosystems and people. It relies on ecological processes, biodiversity and cycles adapted to local conditions, rather than the use of inputs with adverse effects. Organic agriculture combines tradition, innovation and science to benefit the shared environment and promote fair relationships and a good quality of life for all involved"(IFOAM, 1972).

Difference between organic and conventional farming is indicated in Table 3.1.

Table 3.1. Difference between organic and conventional farming

Character	Organic farming	Conventional farming
Nutrient supply	Organic sources like organic manure, crop rotation, cover crop, green manure, nitrogen fixation, bone meal, oil cake and bio-fertiliser application	Chemical fertilisers (NPK) and micro-nutrients
Seed treatment	Natural treatment by using cow urine, soaking and drying	Use of fungicides for seed treatment
Pest control	Natural methods of pest control, use of traps, cultural methods, resistant varieties, pesticides of organic origin	Use of synthetic agro-chemicals like insecticides, fungicides, bactericides etc.
Weed control	Crop rotation, mulching etc.	Use of tillage methods and herbicides
Pollution of environment	No pollution	Causes pollution of soil, water and environment
Food products	Safe for human & animal health	Contains residues of chemicals which are harmful for health of human beings and animals

Principles of organic farming

Organic farming aims at achieving the regeneration and continuance of natural processes of plant growth and development in a given eco-system to make it self-sustainable. The organic sources available in an agro-ecosystem are used for supplying plant nutrients and control of insect pests and diseases. Organic agriculture is based on four basic principles and any system using the methods of organic agriculture and being based on these principles is organic agriculture (Lampkin, N.H, 1994).

The principles of health: Organic agriculture should sustain and enhance the health of the soil, plant, animal, human and planet as one and indivisible.

The principles of ecology: Organic agriculture should be based on living ecological system and cycles, work with them, emulate them and help sustain them.

The principles of fairness: Organic agriculture should build on relationships that ensure fairness with regard to the common environment and life opportunities.

The principles of care: Organic agriculture should be managed in a precautionary and responsible manner to protect the health and well-being of current and future generations and environment.

In organic farming, emphasis is given to supply plant nutrients by promoting nutrient cycles in soil, biological nitrogen fixation, and addition of all possible organic sources of nutrients. Pest and disease management is done by natural enemies, predators, bio-control agents or by use of natural products and cultural techniques Natural balance that exists in the eco-system should not be obstructed by human interference. Although organic farming is based on the above broad principles, actual practice may involve practicing a few of them at a time as ideal conditions may not be available always to follow up all the principles.

Over time, the distinguishing features of organic agriculture came to include, amongst other tenets and practices, the following (Gillman, 2008):

- The maintenance and enhancement of soil fertility through crop rotation, the planting of cover crops and green manures, planting of legumes to fix nitrogen in the soil, and the production and application of compost made from natural materials.
- The development of polycultures (mixed plant communities) rather than monocultures (single crop production) in order to support beneficial insects, soil microorganisms and species diversity.
- Weed management through biological, mechanical and physical means that include regular crop rotation, mowing and tilling, flame weeding, the use of weed suppressing species, mulching and grazing.
- Control of insect pests, fungus and bacteria carried out through interventions such as the introduction of predatory beneficial insects, crop rotation,

companion planting, insect traps, physical barriers, biologic pesticides and herbicides.

- Livestock grazed or fed on organically produced fodder, feed (grain, tubers, hay/silage) and pasture.

Organic farming practices

Organic farming is based on the principles that soil is a living entity and it protects and nurtures the natural micro-organisms in the soil which maintain the soil health without any degradation. Organic farming and food processing practices are socially, ecologically, and economically sustainable food production systems. To be acceptable as organic, crops should be cultivated in lands without any synthetic pesticides, chemical fertilizers, and herbicides for three years before harvesting with enough buffer zone to lower contamination from the adjacent farms. Fertility and nutrient content of soil are managed primarily by farming practices, with crop rotation, and using cover crops that are boosted with animal and plant waste manures. In organic farming use of synthetic fungicides, weedicides, pesticides are prohibited. Natural enemies of the pests are grown and protected to take care of the pests. Bio-fertilizers like Rhizobium, Azotobacter, Azospirillum, PSB and Pseudomonas etc. have been found to be very effective tools of fertility management and biological nutrient mobilization. They are applied as seed treatment, seedling treatment, manure treatment or soil treatment. Many preparations such as liquid manure, Sanjibak, Jeevamrith, Amritpani, Panchagavya etc. are prepared and used for seed treatment, nutrient management and pest management. Quality compost or vermicomposting are used in organic farming to take care of the nutrient need of the crops.

Table 3.2. Various methods of organic farming and potential sources

Purpose	Method	Potential source
Nutrient Management	*In situ* incorporation	Green manure, crop residues, recycling of organic waste, weed biomass & industrial waste
	Application of pre-digested/ semi-digested manures	Animal manures, Farm Yard Manure, Compost, Oil cake, Bone meal, Rock phosphate, Urban compost, cow urine etc.
	Bio-fertilisers	Rhizobium, BGA, Phosphorus solubilising bacteria (PSB), Vermi-compost, Azotobacter, Azospirillum
	Cultural methods	Crop rotation, cover crop, legume cultivation, minimum tillage, agro-forestry, mulching

Pest Management	Cultural methods	Summer ploughing, crop rotation, mixed cropping, inter cropping, use of resistant/ tolerant crop varieties, water management, balanced nutrition
	Biological methods	Encouraging pest defenders, release of predators/parasites, use of NPV, use of herbal products, neem cake and Neem extract, Amritpani, Jibamrith, Panchagavy, Pot manure etc.
	Physical/Mechanical methods	Solar drying, use of traps (yellow, sticky, pheromone), water management and planting trap crops
Soil health management	Promote nutrient cycle, use of organic matter for improving physical, chemical & biological properties of soil without use of agro-chemicals	Minimum tillage, crop rotation, cover crop, application of compost and farm yard manure, oil cakes, mulching,inter cropping, mixed cropping etc.
Water Management	Conserve soil moisture and reduced evaporation	Harvest and conservation of rain water, application of organic manure, contour bunding, terracing, drip irrigation
Weed Management	Prevention and control	Choice of crops, rotation, mulching, inter crop, minimum tillage, biological control
Animal Husbandry	Integrated farming system	Breeding, housing, feeding, fodder cultivation, health coverage and grazing management

Benefits of organic farming

i) **Nutritional benefits:** The organic food is considered better for human health and the environment than conventional production. Organic food is more nutritious and tasty than the products of conventional farms and methods. With lower nitrate content than non-organic varieties organic fruits not only taste sweeter but also have higher antioxidant levels. Organic farming reduces public health risks to farm workers, their families, and consumers by minimizing their exposure to toxic and persistent chemicals on the farm and in food. Evidence shows that food grown organically is rich in nutrients, such as Vitamin C, iron, magnesium, and phosphorus compared to conventionally grown products. Studies have also shown that dairy products from organically raised animals are healthier than conventionally produced dairy products. Animals raised organically are not given antibiotics and are required to be grazed on organically managed pastureland or fed organically grown feed.

ii) **Environmental Benefits**: Organic farmers build healthy soils by nourishing the living component of the soil and the microbial inhabitants by adding soil organic matter that contributes to good soil structure and water-holding capacity. The practices to feed soil biota and build soil organic matter with cover crops, compost, and biologically based soil amendments produce healthy plants that are better able to resist disease and insect predation.

Cover cropping and crop rotation are adopted to change the field ecology, effectively disrupt habitat for weeds, insects, and disease organisms as a result of which they are effectively managed. The wild lives benefit from organic farming and the entire ecosystems and ground water are improved by simply following organic farming methods. Organic farming releases much less carbon dioxide than other farming system. Organic farming has a protective role in environmental conservation as it does not allow synthetic pesticides, most of which are potentially harmful to water, soil, and local terrestrial and aquatic wildlife (Oquist *et al.*, 2007).

iii) **Economic benefits:** By adopting organic farming, conventional farmer can reduce cost of production by 25% as compared to the cost of conventional farming as expensive synthetic fertilizers and pesticides will be eliminated and soil erosion could be minimised. Organic farming practices benefit producer, consumer and dairy animals. Reduction in the use of external inputs and increase in output of organic produces with greater potential to benefit the health of farmers and consumers. More productivity can be ensured through the incorporation of natural process like natural cycles, nitrogen fixation and pest-predator relationship into the agricultural production process. Organic products can fetch premium price in the market and they have export potential.

iv) **Social benefits :** Organic farmers can adopt the practices in small farms and reduce dependency on external inputs and costly technologies thus reducing the competitiveness and disparity among the farmers in a community. It will also lead to food security at the family level and national level. Organic farming is revival of a culture and brings back the indigenous knowledge, beliefs and value system that are almost on extinction now.

v) **Biodiversity conservation:** Crop rotation to build soil fertility and raising animals naturally helps promote biodiversity, which promotes greater health across all living species. When organic farming provides safe environment to wildlife, local ecosystems also improve. Organic farming improves the soil health, avoids risk of groundwater pollution and rehabilitates soil for better conservation of biodiversity. Organic farming encourages healthy biodiversity, which plays a critical role in resilience to adverse weather.

vi) **Climate resilience:** Organic farming reduces the use of non-renewable energy as it reduces or eliminate the use of chemical pesticides and fertilizers, which require large amounts of fossil fuel to produce. Organic farming also returns more carbon to the soil, which then lessens the greenhouse effect and global warming.

vii) **Supporting animal health:** Organic farming not only helps preserve more natural habitat of insects, birds, fish etc. but also encourages birds and other natural predators to live and assists in natural pest control. A clean and chemical-free grazing environment in organic farming helps keep ruminants naturally healthy and resistant to illness. The healthy organic animals produce organic milk or meat.

viii) **Sustainability over the long term:** The concept of sustainable agriculture integrates three main goals-environmental health, economic profitability, and social and economic equity. The concept of sustainability rests on the principle that we must meet the needs of the present without compromising the ability of future generations to meet their own needs. Organic farming is reported to be sustainable as it improves the agro-ecosystem, minimises the overexploitation of natural resources and non-renewable energy, improves the soil health and optimises economic return from farming. Continuous degradation of soil fertility by chemical fertilizers leads to production loss and hence increases the cost of production which makes the farming economically unsustainable.

Soil building practices such as crop rotations, inter-cropping, symbiotic associations, cover crops, organic fertilizers and minimum tillage are central to organic practices. These encourage soil fauna and flora, improving soil formation and structure and creating more stable systems. In turn, nutrient and energy cycling is increased and the retentive abilities of the soil for nutrients and water are enhanced, compensating for the non-use of mineral fertilizers. Such management techniques also play an important role in soil erosion control. Well managed organic systems with better nutrient retentive abilities, greatly reduce the risk of groundwater pollution. Organic agriculture contributes to mitigating the greenhouse effect and global warming through its ability to sequester carbon in the soil. Many management practices used by organic agriculture (e.g. minimum tillage, returning crop residues to the soil, the use of cover crops and rotations, and the greater integration of nitrogen-fixing legumes), increase the return of carbon to the soil, raising productivity and favouring carbon storage.

ix) **Ecological services**: Ecological services such as soil forming and conditioning, soil stabilization, waste recycling, carbon sequestration, nutrients cycling, predation, pollination and habitats are favoured by organic farming. Not only does organic farming build healthy soil, but it helps combat serious soil and land issues, such as erosion.

Limitations of organic farming

i) **Food security:** There is apprehension that organic farming may not be able to meet the future demand for food as the population goes on increasing and we have to increase the yield substantially.In organic farming yield increase and higher return cannot be expected in a short time and there will be gradual increase in yield after three to eight years after stabilisation of organic carbon in the soil. Usually the high yielding varieties are not suitable for organic farming as they require more of fertilisers and pesticides for expression of their yield potential.

ii) **Availability of organic manure:** With the decline of animal population it may be difficult to arrange required quantity of organic manure for organic farming. There are some recommended biological preparations, which can be produced by trained farmers. Hence organic farming has become more knowledge intensive and labour intensive.

iii) **Package of practices**: Uniform package of practices for organic farming may not be available for all locations. Taking the availability of organic manures into consideration, separate package of practices have to be developed for different agro-ecosystems.

iv) **Certification**: Organic products require certification before marketing. Since the procedure is cumbersome and cost involving, the smallholders face difficulty in getting their product certified. For this, they do not get premium price in the market.

Status of organic farming

Worldwide Organic producers by Region 2017

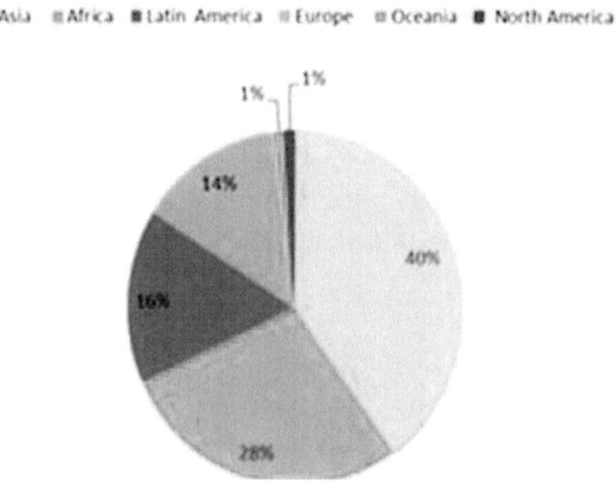

Fig. 3.1. Status of organic farming

Since 1985, the total area of farmland under organic production has been increased steadily over the last three decades (Willer and Lernoud, 2019). Asia contributes to the largest percentage (40%) of organic production in the World with largest number of organic farmers.

Government of India has taken steps for promotion of organic farming in the country through implementation of the following schemes/programmes:

- The Paramparagat Krishi Vikas Yojana (PKVY);
- Organic Value Chain Development in North Eastern Region Scheme;
- Rashtriya Krishi Vikas Yojana (RKVY);
- The Mission for Integrated Development of Horticulture (MIDH);
- National Programme for Organic Production;
- National Project on Organic Farming; and
- National Mission for Sustainable Agriculture (NMSA)

The global organic food market has been reached USD 81.6 billion in 2015 from USD 17.9 billion during the year 2000 and most of which showed double-digit growth rates (Willer and Lernoud, 2019).

3.3.2 Ecological farming

Ecological Farming is a production system based on the principles of agro-ecology. It is diverse, ecologically sound and viable. In spite of many similarities, ecological farming is not the same as organic farming. It includes all methods, which regenerate ecosystem services like prevention of soil erosion, water infiltration and retention, carbon sequestration in the form of humus, and increased bio-diversity. The techniques are used including no-till, multispecies cover crops, strip cropping, terrace cultivation, shelter belts, pasture cropping etc. By using biological processes in place of off-farm inputs, ecological farming aims to improve soil conditions, diversify species and genetic resources, enhance beneficial biological interactions, recycle biomass, and minimize energy loss (Lee, 2005). There are many types of ecological farming such as Natural Farming, Permaculture, Zero Budget Natural farming, Bio-dynamic agriculture, Regenerative agriculture, and Homa farming.

A summary of the distinguishing characteristics of ecological and conventional farming appears in Table 3.3.

Table 3.3. Characteristics of ecological and conventional agriculture

Characters	Ecological farming	Conventional farming
Approach	Holistic approach	Target approach
Techniques	Poly-cropping, intercropping, cover crops, mulching, crop rotation, minimum tillage, use of natural parasitic relationships, construction of rainfed or drip irrigation systems	Mono-cropping, application of chemical fertilizers and pesticides, construction of irrigation systems, use of hybrid and transgenic seed varieties
Distinctive inputs	Intensive labour; access to manure or other organic fertilizers; education and training in ecological farming practices; adaptation of practices to local context	Physical and financial access to mineral fertilizers, synthetic pesticides, and improved crop varieties; irrigation sources or adequate rainfall
Goals	Sustain and enhance food production, environmental conditions, and social relationships in perpetuity	Increase yields and productivity

Sometimes poor farmers cannot afford for external inputs which are seen as organic by default. Such farming practices due to poverty or lack of resources cannot be called ecological or organic farming.

Principles of ecological farming

Ecological farming applies ecological principles and approaches to agricultural ecosystems. An ecological approach involves designing the strengths of natural ecosystem into agro-ecosystems. Natural ecosystems have efficiency, diversity, self-sufficiency, self-regulation and resilience.

Efficiency	In a natural ecosystem sun's energy is captured by green plants to produce food for consumers at different levels which are transmitted by food-chains and food-webs. Natural ecosystem is efficient in capturing rainfall and mobilising nutrient cycles.It also maintains ground and surface water and reduces nutrient losses.
Diversity	Biological diversity above the soil and in the soil provides a check and balance for nutrient availability and control of pest and diseases in a natural way.
Self-sufficiency	The natural ecosystem is self-sufficient in its requirement depending on sunlight and precipitation.
Self-regulation	There is a natural defence mechanism between insect pests and natural defenders in the ecosystem.
Resilience	There are many disturbances in the natural ecosystem and the stronger ones become resilient to such disturbance and are capable of recovery.

Ecological farming approaches

The overall goals of ecological farming are to create conditions in soil and above-ground that encourage the growth of healthy plants, and control pests by the natural defenders. This is a sort of 'habitat management and conservation' in soil and above ground to enhance biological diversity and growth of healthy plans in space and time. The approaches are summarised below.

Diversity: Nature's diversity protects natural habitat and biodiversity along with encouraging seed diversity, nutrient cycle, regeneration of soil and maintaining natural balance between crop pests and natural enemies.

Technology: Both the modern and indigenous technologies are combined to grow crops with diverse varieties under climate change without loss of bio-diversity and use of agro-chemicals.

Soil health: Ecological farming maintains the soil health by protecting soil erosion, pollution and other problems by management of soil organic matter, water and prevention of land degradation. Addition of organic matter improves the physical, chemical and biological property of soil.

Environment protection: Due to elimination of agro-chemicals the nearby water bodies are not polluted and thus emission of greenhouse gas (GHG) is reduced.

Practices adopted in ecological farming

Agro-ecological techniques encompass a range of natural practices such as crop diversification; cover cropping, crop rotation, integrated farming etc. The common practices adopted in different types of ecological farming are summarised below.

Crop diversity: Discouraging monoculture in ecological farming and adopting polyculture along with crop rotation. Modification of planting pattern like appropriate crop geometry, trap crops, border crops to attract beneficial insects, cover crops, crop rotation with legumes, mixed and inter cropping etc.;

Seed diversity: Selection of varieties resistant to local crop pests with high yield potential and prohibition of the use of genetically modified organisms;

Soil fertility management: Addition of large quantities of organic matter on regular basis for maintenance of soil health for growth of stronger plants that can resist pest attack and weather aberrations. Organic matter can have positive impact on nutrient availability, maintaining a good soil-air-water relationship as a result of which there will be good conditions for plant roots to grow and explore, production of plant growth stimulating compounds by microorganisms, disease suppression, low weed seed populations, etc.

Pest management: Choosing different plant species and animal breeds that are better suited to local conditions and potential problems with pests and diseases. Proper soil, water and nutrient management without application of pesticides will encourage the activities of natural defenders to control insect pests in crop fields. Healthy plants created by better soil & above ground conditions will enhance defences against pests.

Improvement of soil properties: Reduced tillage and mulching are required for control of weeds, maintenance of soil structure and controlling soil erosion. Soil covered with mulch (living vegetation and/or crop residues by using cover crops, sod crops in rotations, and/or reduced tillage practices) to encourage water to infiltrate into the soil instead of running off the field. Minimum tillage or zero tillage reduces soil compaction to a minimum by keeping heavy machinery away.

Integrated farming system: Encourage livestock farming or integrated farming system for sustainability in agriculture.

Benefits of ecological farming

Ecological Farming strengthens our agriculture, and effectively adapts our food system to changing climatic conditions and economic realities that create resilience. Growing different crops at the field levels is a proven and highly reliable way to make our agriculture resilient to increasingly unpredictable changes in the climate. The benefits of ecological farming are:

- With large amounts of organic matter in the soil, there will be improvement is physical, chemical and biological properties of soil which would provide a better condition for plant growth.
- Nutrient cycling, biological nitrogen fixation, and soil regeneration would reduce carbon emissions. Integration of livestock plays a key role in agro-ecosystems.
- The agro-products of ecological farming are safe, nutritious and free from any contamination which promotes health of human beings and animals.
- The soil rich in organic matter can improve the water holding capacity of soil, improve the rate of infiltration, reduce soil erosion and resist the climate induced adversities like drought and flood.
- Ecological farming will emit limited amount of greenhouse gas to the atmosphere as a result of which contribution of agriculture to climate change will be reduced to minimum.

TYPES OF ECOLOGICAL FARMING

3.3.2.1 Natural farming

Natural farming is a system of farming, where the cultivation is left to the natural processes with least human interference. It is a sort of primitive agriculture

Fig. 3.2. Fukuoka M. and natural farming

which believes that nature can take care of crops. "The One-straw Revolution" written by Fukuoka, M (1975) interested many Japanese farmers for natural farming. He conceptualised "do nothing" agriculture method and go with nature. The four principles of natural farming, according to him, were (i) no cultivation, (ii) no chemical fertilisers and prepared composts, (iii) no weeding by tillage or herbicides, and (iv) no dependence on chemicals comply with the natural order and lead to the replenishment of nature's richness. Spreading un-cut straw as mulch was considered as connected with everything, with fertility, with germination, with weeds, with keeping away sparrows, and with water management. Growing rice in a dry field was demonstrated by such type of farming.

Sir Albert Howard, in his book "An Agricultural Testament" (1943) has stated "maintenance of the fertility of the soil is the first condition of any permanent system of agriculture". Nature's agriculture seen in forests indicates mixed farming and natural way of soil management. In this type of natural farming soil is protected from direct exposure of sun, rain and wind as a result of which soil health is preserved and erosion does not take place. There is mutual interaction between crops and livestock. Sir Albert Howard has summed up the natural farming as " Mother earth never attempts to farm without live stock; she always raises mixed crops; great pains are taken to preserve the soil and to prevent erosion; the mixed vegetable and animal wastes are converted into humus; there is no waste; the processes of growth and the processes of decay balance one another; ample provision is made to maintain large reserves of fertility; the greatest care is taken to store the rainfall; both plants and animals are left to protect themselves against disease".

Although there are some similarities between organic farming and natural farming, still there are many differences in both the systems as summarised below (Table 3.4).

Table 3.4. Organic farming and Natural farming

Particulars	Organic farming	Natural farming
Agricultural practices	Organic farming requires basic practices like ploughing, tilling, mixing of manures, weeding, etc.	In natural farming there is no ploughing, no tilting of soil and no fertilizers, and no weeding is done as in natural ecosystems
Chemical use	Organic fertilizers and manures like compost, vermi-compost	Neither chemical nor organic fertilizers are added to the soil.

	and cow dung manure are used and added to crop fields.	
Cost	It is still expensive due to application of bulky organic manures.	It is an extremely low-cost farming method, relying on local ecological impact on the biodiversity.
Efforts	In organic farming tilling is done and the manures and composts are incorporated into the soil for their proper decomposition and this requires more effort.	Here decomposition of organic matter by microbes and earthworms is encouraged right on the soil surface itself, which gradually adds nutrition in the soil.
Effect	Organic farming has slight adverse effect on the surrounding environment as it involves intervening with the natural cycles.	Natural farming does not have any effect on the surrounding environment and it conforms with local the natural processes.
	Organic farming is more about using naturally available resources optimally to enhance productivity and production.	Natural Farming is minimalist in operations and human intervention, leaving things to nature to manage.

Principles

Natural farming is a system that makes all inputs from natural materials within laws of nature. The soil and water become clean where natural farming is practiced. It opposes human exploitation and respects the nature of life. Similarly healthy animals are reared rather than feeding them with hormones and antibiotics. The farm produce from natural farming are of high quality and good taste without any contamination or toxic residues. These products have comparatively more protein, amino acid, crude fat and other essential nutrients than the conventional products and chemical residue such as nitrate is almost undetectable. Bases on nutritive theory cycle, farmers produce nutrients and improve soil conditions and pests are controlled naturally. This is the reason why it can be effective means for resource poor farmers who cannot buy external inputs like fertilisers and pesticides.

Natural farming methods include "No-tillage, No weeding, No chemical fertilisers and pesticides".

No-tillage: Soil is not tilled in natural farming mechanically and the job is left to the earthworms. Tillage cultivates the soil to a shallow depth of 20 cm while earthworm can cultivate up to 7 m deep. The soil productivity increases due to excretions of the earthworm and surface weeds are controlled by no-tillage.

No-fertilisers: No fertiliser or prepared compost is used in natural farming. The essential nutrients like nitrogen, phosphorous, potassium, calcium etc. are not applied from external sources and substituted with natural farming inputs such as fish, egg shells, bone meals etc. in addition to biological nitrogen fixation.

No pesticide and herbicide: Pesticides kill both pests and natural enemies and leave residues in soil, water and farm produce. Instead of using such toxic chemicals, natural farming uses natural methods of pest control by creating an ecology that has recovered the natural balance between the pests and defenders to reduce pests and disease occurrence. Similarly, alternate methods like cover cropping, mulching and crop rotation are used for control of weeds rather than using herbicides. Mulching gives excellent result for growth of soil organisms like earthworm and keeps the soil temperature, moisture and weed under control. Natural orchards are green with grass growing on the interspace of fruit trees which prevent soil erosion, hold moisture, propagate soil micro-organisms and improve the soil condition.

No pollution: Under natural farming the animals do not emit wastewater and rice straw, sawdust, fresh soil are used for flooring which enables quick decomposition of faeces on the floor. The floor remains dry and fluffy by movement of sunlight, wing and micro-organisms. In absence of chemical fertilisers and pesticides, there is no problem of soil and water pollution.

Nutritive cycle theory: Natural farming follows the scientific approach of just application of nutrients using the right material at the right amount and at the right stage for optimal growth of crops and livestock. The Nutritive Cycle Theory states that plants and livestock need different nutrients during different stages of growth which guides farmers in proper nutrient management.

Practices of natural farming

Mulching: Mulching is a practice of soil cover with crop residues or other materials to check evaporation and maintain the soil condition for germination and crop growth. They control weeds and supply nutrients on decomposition.

Green manuring: Crops like *dhaincha* (*Sesbania aculeata*) or sannhemp is sown (usually) just before the monsoons and incorporated into soil just around flowering (30-45 days after sowing) after which the season's main crop is planted. Green manuring adds nitrogen by biological fixation and supply biomass to improve soil properties.

Cover cropping: Soil cover with nitrogen-fixing crops that grow fast is done for covering the soil in the fallow months. This adds nitrogen to the soil, suppresses weeds, prevents soil erosion and adds biomass to soil.

Biological / natural pest and weed control

In an agricultural eco-system, parasites and predators can control the harmful pests. When deliberate use of chemicals is ensured, the natural defenders are killed for which the pests increase in number in the absence of any natural check. In natural farming the farmers use prophylactics such as diluted cow urine and vermi-wash (the fluid from a vermi-compost tank) both of which can also be used in greater strengths as pesticides. Integrated pest control is done by mechanical control, use of various traps (pheromone, sticky, light), growing trap crops, use of bio-agents and organic pesticides, neem, ginger, chill, *Vitex negundo* (Indian pivet tree), custard apple (the seeds), *Pongamia pinnata* (pongammia/karanj), asafoetida, turmeric, garlic, tobacco, sweet flag, *Nux vomica*, tulsi etc. Each pest requires a specific preparation.

Composting : The organic wastes can be properly recycled to form excellent compost in one to six months, depending upon the composting process used. Every farm can choose or even develop a suitable compost process depending upon its own needs and resources, including availability of labour, managerial time and investment potential. One method of composting farm wastes is by vermicomposting which uses earthworms to eat and break up the organic wastes. There are a number of other methods and innovations, adaptations and improvements are always possible. Methods can be aerobic or anaerobic and above ground or below, though the best way to get high quality compost quickly is to make a heap above the ground. NADEP method of composting is now widely adopted to prepare compost within a short time.

Polyculture: Monoculture leads to many environmental problems. If a single variety is widely grown, a pest or disease to which it lacks resistance can lead to a dramatic fall in production. Crop diversification, including rotation and intercropping and the use of diverse forage plants in pastureland, can reduce pest damage and weed invasions.Traditional farmers follow the systems of crop rotation, multi-cropping, inter-cropping and polyculture to make maximum use of all inputs available to them, including soil, water and light, at a minimum cost to the environment.

Crop rotation: Crop rotation is the sequence of cropping where two dissimilar types of crops follow each other in a rotation. It reduces reliance on one set of nutrients, pest and weed pressure, and the probability of developing resistant pest and weeds.

Multi-cropping: Multiple cropping is a practice of growing two or more crops together in a farm. An example of multi-cropping is tomatoes + onions + marigold (where the marigolds repel some of tomato's pests). Inter-cropping is the cultivation of another crop in the spaces available between the main crop.

A good example is the multi-tier system of coconut + banana + pineapple/ginger/ leguminous fodder/medicinal or aromatic plants.

3.3.2.2 Zero Budget Natural Farming (ZBNF)

Of the many working models of natural farming all over the world, the Zero Budget Natural Farming (ZBNF) evolved by Padmshri Subash Palekar in India is most popular. It is a unique method which requires no monetary investment for purchased inputs like fertilisers and pesticides from the market. The farmer grows local varieties of crops without any budget for chemicals. The system requires native breed of cattle and it is claimed that one cow is sufficient to take up this method of farming on thirty acres of land.

In ZBNF the farmer can produce his own seed or he may use seeds that are available locally which is grown without application of chemical fertilisers and pesticides. The requirement of hired labourer is also reduced to the bare minimum as the system discourages intercultural operations. It is most suitable for resource poor farmers to grow their crops with minimum cost. It is believed by the practitioners that 1.5 to 2.0% of nutrient from soil is used for crop production and remaining 98 to 98.5% nutrients are taken from air, water & solar energy. It is further believed that the green leaves are food producing factories. The leaves take carbon dioxide & nitrogen from the air, water from the canal, river or well given by the monsoon clouds, and solar energy from the sun for producing the food. Every green leaf of any plant produces 4.5 gram carbohydrates per square feet surface, from which we get 1.5 gram grains or 2.25 gram fruits.

Four wheels of ZBNF

Four Pillars of Zero Budgt Natural Farming

Fig. 3.3. Four pillars of ZBNF

BIJAMRITA (Seed treatment using local cow dung and cow urine), JEEVAMRITHA (applying inoculation made of local cow dung and cow urine without any fertilizers and pesticides), Mulching (activities to ensure favourable micro - climate in the soil) and Waaphasa (soil aeration) are the four pillars of ZBNF.

a) BIJAMRITA: It is prepared by using cow urine, dung and soil for seed treatment before planting. In modern farming a number of fungicides and insecticides have been recommended for treating the seeds before sowing which also destroy all the useful and effective microorganisms. When the seeds treated with these poisonous chemicals germinate and grow, these poisons are also absorbed by the roots with the soil water solution and are deposited in the body organs of the plant i.e. vegetables, grains, fruits, tubers etc. which affect our health after eating. Under ZBNF the seeds are treated with Bijamrita for checking seed borne diseases.

Fig. 3.4. Preparation of Bijamrith

Preparation method

Materials used: 20 litre water, 5 Kg local cow dung, 5 litre local cow urine, 50 g lime & small quantity of soil from the bund of the farm.

Method: Take 5 Kg local cow dung in a cloth and bound it by tape. Hang this in the 20 Litre water up to 12 hours. Take one litre water and add 50 g lime in it and let it stabilize for a night. Then next morning, squeeze this bundle of the cow dung in that water thrice continuously, so that all essence of cow dung will accumulate in that water. Then add a handful of soil in that water solution and stir it well. Then add 5 litre of desi cow urine in that solution & add lime water and stir it well. Bijamrita is then used to treat the seeds by spreading it over the seeds, mixing these seeds by hands, drying it well and use for sowing.

b) JEEVAMRITHA: 'Jeevamrutha' promotes immense biological activity in the soil and makes the nutrients available to the crop. Jeevamrutha is applied once in a fortnight as nutrient which also acts as a catalytic agent to promote biological activity in the soil.

   Jeevamritha also helps to prevent Jeevamritha fungal and bacterial plant diseases. Jeevamrutha is only needed for the first 3 years of the transition, after which the system becomes self – sustaining.

c) MULCHING (ACCHADANA): For the proper growth, multiplication and activity of beneficial micro-organisms that are applied through Jeevamritha, a favourable definite microclimate is required. In this favourable microclimate the temperature of the soil should be in the range of 25 to 32°C with 65 to 72% moisture, darkness and warmth. Mulching creates such micro-climate. There are three types of mulching such as soil mulching, straw mulching and live mulching. The cultivation of the field is considered as soil mulch, which circulates the air in the soil and intercepts the raindrops to flow and to conserve them in the soil . Conserved rain water storage is essential for the growth of the crops and the stoppage of the rainwater flow to restrict the topsoil erosion and control weeds. But cultivation should be limited to the soil layer up to 10-15 cm, in which, these feeding roots and micro-organisms are active.

Application of dried straw biomass of the previous crops as a soil cover in the succeeding crop is called straw mulching which is very important as the seeds are covered by this straw mulch and it saves seeds from birds, insects and animals. It also creates a micro-climate which activates micro-organisms and local earthworms. It creates a favourable condition for decomposition of organic matter in soil such as roots to prepare humus in the soil for future new crop. Straw mulch conserves soil moisture in the soil and reduces the evaporative loss of water from the soil making it available for the utilization by the soil micro-organisms, besides moderating soil temperature and protecting humus from extreme temperatures.

Live mulching by intercrops and mixed crops has a symbiotic association with each other. Mostly legumes are grown as intercrops as they help in fixing atmospheric nitrogen and make it available to the main crop. Similarly crops having different requirements of light, space and nutrients are grown together to avoid competition. Similarly the mixed crop pattern of monocot in dicot & dicot in monocot helps to supply the essential elements to the crops. The dicots supply nitrogen by means of nitrogen fixing bacteria and monocots supply other elements like potash, phosphate, sulphur etc.

d) WAAPHASA (SOIL AERATION): Water is the life blood of soil. If there is Waaphasa in the soil, the water is life. If there is no Waaphasa in the soil, water is not stored for the plant and soil biota. Waaphasa is that microclimate in the soil, by which the soil organisms and roots can live freely with availability of sufficient air and essential moisture in the soil. Waaphasa means the mixture of 50% air and 50% water vapours in the cavities between two soil particles which has to be maintained in the top 10-15 cm of soil that is the active root zone for the micro-organisms that facilitate absorption of water and nutrients.

Other important principles of ZBNF: Intercropping is an important practice in ZBNF in which there is a close association between the crops and trees growing on the farm. There is need for contour bunds for preservation of rain water which should be efficiently used to enhance the productivity and water use efficiency of different crops. Local species of earthworms through increased organic matter is recommended instead of use of *Eisenia foetida* only feeds on the organic biomass and does not eat and burrow the soil. *Eisenia foetida* feed on organic matter, as the result, the mulching of the organic matter on the soil is totally destroyed. They also accumulate large quantity of heavy metals in their bodies which is further transferred to the soil in the form that can be easily absorbed by the crop. Cow dung from the local cow is most beneficial and has the highest concentrations of micro-organisms as compared to European cow breeds such as Jersey and Holstein. Palekar says "cow dung should be used as fresh as possible while the urine used should be as old as possible as they are more effective".

Zero Budget Natural Farming advocates cultivation of diverse species of crops depending on site specific agro climatic conditions. Mixed cropping provides buffer against total failure of single crop and also widens the income source of farmers. Crop rotation is also emphasized to discourage build-up of endemic pests. In the scheme of mixed cropping, cereals, millets, leguminous crops, horticulture crops particularly vegetables and even medicinal plants can be included to make farming more lucrative.

Plant protection: The plant protection agents that can be used by farmers in ZBNF include. Agniastra, Brahmastra and Neemastra which are home-made preparations.

AGNIASTRA: Agniastra is a very powerful agent against pests like leaf roller, stem borer, fruit borer & pod borer. The solution can be prepared in-house by the farmers. Cow urine (go - mutra), crushed leaves of tobacco @ 25 g per litre of cow urine, deshi hot green chilli pulp @ 25 g per litre of cow urine, deshi garlic pulp @ 12.5 g per litre of cow urine, crushed neem leaves (with thin stems) or neem seed powder @ 100 g per litre of cow urine are mixed thoroughly in an earthen pot for preparation of Agniastra. Use wooden stick for mixing the ingredients. The stick should be moved clockwise while mixing so that positive energy is circulated in the mixture. Boil the solution and cover the pot with gunny bag or poly net. Let the mixture ferment for 48 hours. Stir the solution clockwise using a wooden stick 2 times a day for 1 min. After 48 hours filter the solution and store Agniastra in bottle.

The solution should be sprayed as foliar spray using 3% Agniastra with water. If the infestation is high then you can use 4% solution. For 1 acre mix 6 to 8 litres of Agniastra with 200 litres of water and spray on the plants.

BRAMHASTRA: Crush 3 kg neem leaves in 10 litre cow urine. Crush 2 kg custard apple leaf, 2 kg papaya leaf, 2 kg pomegranate leaves, 2 kg guava leaves in water. Mix the two and boil 5 times at some interval till it become half. Keep for 24 hours, then filter and squeeze the extract. This can be stored in bottles for 6 months. This preparation is highly effective against sucking pests, pod / fruit borers. For spraying the stock solution can be diluted by adding 2.0 - 2.5 litre of this extract in 100 litre water and sprayed over 1 acre.

NEEMASTRA: Take 100 litre water and add 5 litre local cow urine and 5 Kg local cow dung in it. Crush 5 Kg of neem leaves & add this neem pulp in this water. Let this solution ferment for 24 hours. Stir this solution twice a day by any stick. Filter this by cloth. Spray this Neemastra as it is on the plants for sucking pests & mealy bug. This preparation is effective against white fly, jassids, aphids etc.

3.3.2.3 Permaculture

Permaculture originally came from "permanent agriculture" but was later adjusted to mean "permanent culture", incorporating social aspects. The term was coined in 1978 by Bill Mollison and David Holmgren, who formulated the concept in opposition to Western industrialized methods and in congruence with Indigenous or traditional knowledge. Mollison and Holmgren (1978) described permaculture as, "an integrated, evolving system of perennial or self-perpetuating plant and animal species useful to man". In words of Mollison, permaculture is a "conscious design and maintenance of agriculturally productive ecosystems which have the diversity, stability and resilience of natural ecosystems. Holmgren later expanded the definition to, "consciously designed landscapes which mimic the patterns and relationships found in nature, while yielding an abundance of food, fibre and energy for provision of local needs" (Holmgren, 2003). A permaculture system is a system that resembles nature and is based on natural cycles and ecosystems (Holzer, 2004). It is the harmonious integration of the landscape with people providing their food, energy, shelter and other material and non-material needs in a sustainable way (Bell, 2004). Management of carbon in holistically managed land and conventionally managed land is indicated in Fig. 3.5.

Permaculture is a multidisciplinary toolbox that integrates agriculture, water harvesting and hydrology, energy, natural building, forestry, waste management, animal systems, aquaculture, appropriate technology, economics and community development. The philosophy behind permaculture is working with nature rather than against the nature. The difference between the cultivated ecosystem and natural ecosystem is the use of great majority of species for the humans and livestock. Recycling of nutrients and energy in nature is a function of many species.

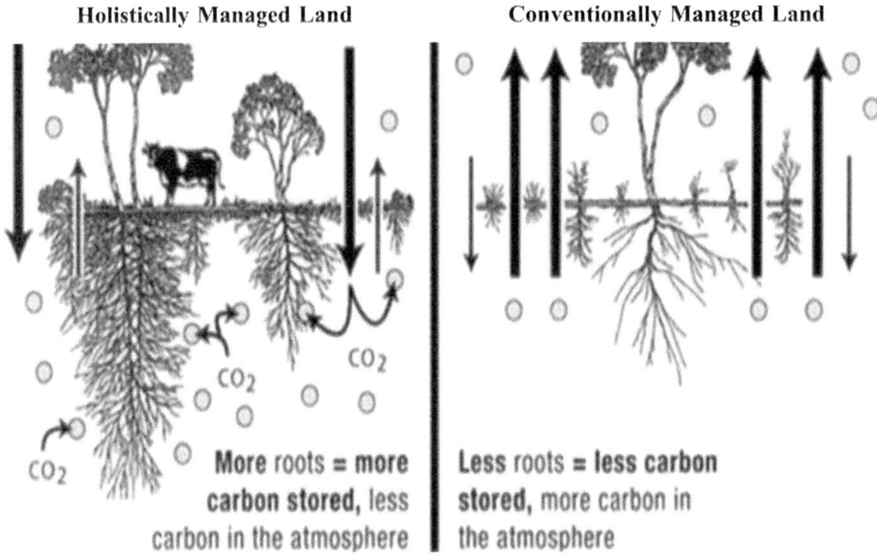

Fig. 3.5. Carbon management in natural ecosystem

Some key aspects of permaculture include

- The three core tenets of permaculture are care for the people, care for the earth and return of surplus.
- Permaculture systems strive to maximise the layers (canopy, understory, ground cover) (Mollison, 1988).
- Permaculture takes into account the interactions in the system between the human environment and activity, and the different species that dwell in that system (Burnett, 2001).

Permaculture has been implemented and has gained widespread visibility throughout the world as an agricultural and architectural design system and as a guiding life principle or philosophy. Traditional and indigenous practices are highly valued in permaculture because they have been developed in perpetual dialogue with specific climate and soil conditions.

Holmgren (2013) articulated twelve permaculture design principles in his Permaculture: Principles and pathways beyond sustainability:

- Observe and interact: Take time to engage with nature to design solutions that suit a particular situation.
- Catch and store energy: Develop systems that collect resources at peak abundance for use in times of need.
- Obtain a yield: Emphasize projects that generate meaningful rewards.

- Apply self-regulation and accept feedback: Discourage inappropriate activity to ensure that systems function well.
- Use and value renewable resources and services: Make the best use of nature's abundance: reduce consumption and dependence on non-renewable resources.
- Produce no waste: Value and employ all available resources: waste nothing.
- Design from patterns to details: Observe patterns in nature and society and use them to inform designs, later adding details.
- Integrate rather than segregate: Proper designs allow relationships to develop between design elements, allowing them to work together to support each other.
- Use small and slow solutions: Small and slow systems are easier to maintain, make better use of local resources, and produce more sustainable outcomes.
- Use and value diversity: Diversity reduces system-level vulnerability to threats and fully exploits its environment.
- Use edges and value the marginal: The border between things is where the most interesting events take place. These are often the system's most valuable, diverse, and productive elements.
- Creatively use and respond to change: A positive impact on inevitable change comes from careful observation, followed by well-timed intervention.

Ethics of permaculture

The ethics on which permaculture builds as stated by Mollison (1988) are:

- "Care of the Earth: Provision for all life systems to continue and multiply".
- "Care of people: Provision for people to access those resources necessary for their existence"
- "Setting limits to population and consumption: By governing our own needs, we can set resources aside to further the above principles"

Common practices of permaculture

i) **Agroforestry :** Agroforestry is a farming system that integrates trees and shrubs with crops or livestock. It utilises both agricultural and forestry technologies to create more diverse, productive, profitable, healthy and sustainable land-use systems.

ii) **Urban permaculture:** Urban permaculture aims at utilisation of space for food production and minimisation of waste.

iii) **Vermicomposting**: Vermicomposting is an improved method of composting using earthworms to break down green and brown waste. The worms produce worm castings, which can be used to organically fertilize the garden. Worm castings have been noted to increase plant growth and decrease heavy metals in the soil.

iv) **Natural building**: It involves using a range of building systems and materials that apply permaculture principles. Natural building attempts to lessen environmental impacts of buildings without sacrificing comfort, health, or aesthetics. Natural materials (e.g., clay, rock, sand, straw, wood, reeds), and draws heavily on traditional architectural strategies found in various climates.

v) **Rainwater harvesting** : It is the accumulation and storage of rainwater for reuse before it runs off or reaches the aquifer (Aramo, 2012). The harvested rainwater is used as drinking water, water for livestock, and water for irrigation, as well as other domestic and industrial uses. Rainwater collected from the roofs of houses and local institutions can make an important contribution to the availability of drinking water. It can supplement the water table and increase urban greenery. Grey water is wastewater generated from domestic activities such as laundry, dishwashing, and bathing, which can be recycled for uses such as landscape irrigation and constructed wetlands. Grey water is largely sterile, but not potable (drinkable).

vi) **Domesticated animals:** Domesticated animals are a critical component of any sustainable ecosystem. Activities that contribute to the system include: foraging to cycle nutrients, clearing fallen fruit, weed maintenance, spreading seeds, and pest maintenance. Nutrients are cycled by animals, transformed from their less digestible form (such as grass or twigs) into more nutrient-dense manure.

vii) **Sheet mulching:** Mulch material includes stones, leaves, cardboard, wood chips, and gravel, although in permaculture mulches of organic material are preferred because they perform more functions. Mulching helps absorbing rainfall, reducing evaporation, providing nutrients, increasing soil organic matter, creating habitat for soil organisms, suppressing weed growth and seed germination, moderating diurnal temperature swings, protecting against frost, and reducing erosion. Sheet mulching mimics the leaf cover that is found on forest floors. Sheet mulch serves as a "nutrient bank," storing nutrients contained in organic matter and slowly making these nutrients available to plants as the organic matter slowly and naturally breaks down. It also improves the soil by attracting and feeding earthworms, and many other soil micro-organisms, as well as adding humus.

viii) **Conservation grazing**: Grazing is blamed for much destruction. However, controlled grazing is not harmful. In conservation grazing the primary purpose of the animals is to benefit the environment and the animals are not necessarily used for meat, milk or fibre

ix) **Fruit tree management:** Many workers suggest that trees should be allowed to grow un-pruned.

In summary it can be said that permaculture is more than food production – it is a design process that can be applied to organizations, homes, and landscapes. Permaculture's three ethics succinctly present the roots that hold together all that has been presented are: care for the earth, care for the people, and there are limits to growth (Holmgren, 2007). To care for the earth and people, and to recognize limits to growth, is to realize our need for regeneration. Regeneration of food and landscapes, as opposed to degeneration, is a necessary standard for environmental sustainability and applied permaculture.

3.3.2.4 Bio-dynamic agriculture

Bio-dynamic farming involves use of cosmic rays and cosmic energy of nature which could be used for enriching natural sources of nutrients. Bio-dynamic agriculture is a form of alternative agriculture which treats the farm as a living organism. The 'biological-dynamic' approach, based on Steiner's indications for considering the spiritual foundations of agriculture, came in time to be known as bio-dynamics and it goes under that name to this day. Bio-dynamics agriculture conceives the farm an 'organism', 'individuality', or 'self-contained entity' (Steiner, 1993) with, as much as is practically possible, the needs of the various members of the farm organism being met from within the farm itself (nutrient needs, re-incorporation of waste products into the fertility cycle, careful conservation of energy and nutrient flows). This principle is proposed in order to develop resilience and long-term sustainability for the whole farm (Masson, 2012; Thornton Smith, 2009; Hurter, 2014). Attention is placed by biodynamic practitioners on supporting "earthly and cosmic forces that form life" (Koepf, 1989).

Biodynamic practice hinges on the notion of vitality and on supporting/enhancing life processes in soil, plant, animal, and human. Already in 1924 Steiner drew attention to the importance of soil fertility and proposed that a healthy soil (implying soil life) was the basis for healthy plants and animals and, ultimately, was the basis for healthy food for human consumption (Steiner, 1993; Pfeiffer, 1947/2004). It is precisely the production and use of these herbal 'preparations' (coupled with the adherence to astronomical planting calendars) that has led several critics of bio-dynamics to dismiss it as an approach founded on pseudoscience and/or 'magical thinking' (Holger, 1994; Chalker-Scott, 2004).

An alternative would be to take these preparations (and the planting calendar used by biodynamic practitioners) as a stimulus for a study of paradigmatic differences in agricultural practices (Lorand, 1996) and to research not only what a given approach entails as practices but how these arose from an epistemological and ontological perspective (Code, 2014). The preparations are, in this light, one aspect of an approach to agriculture (i.e. biological-dynamic) that arises from a very different worldview and understanding of the organisms, beings and processes in nature than is evident in either conventional agriculture or, in fact, in other 'alternatives' to conventional agriculture (Code, 2014; Thornton Smith, 2009).

Methods unique to the biodynamic approach include its treatment of animals, crops, and soil as a single system, an emphasis from its beginnings on local production and distribution systems, its use of traditional and development of new local breeds and varieties. Some methods use an astrological sowing and planting calendar. In bio-dynamic farming the farm is conceived as an organism and emphasis is placed on integration of crops and livestock, recycling of nutrients, maintenance of soil health and wellbeing of the crops and animals. In this system food is raised bio-dynamically with better quality than the conventional products.

Bio-dynamic preparations

Biodynamic farming uses nine biodynamic preparations described by Steiner (1924) for the purpose of enhancing soil quality and stimulating plant life. There are nine BD preparations which are used in bio-dynamic farming as indicated in Table 3.5. Biodynamic compost prepared from BD 502-507 serves as a way to recycle animal manures and organic wastes, stabilize nitrogen, and build soil humus and enhance soil health.

Table 3.5. Preparation of BDF products

BD preparation	Method of preparation	Method of use
500 : Cow horn manure	• Feed cattle with high quality food for two days prior to collecting dung for BD 500 • Prepare burial pit of 18" deep having good quality earth without any growth or flooding. • Collect clean cow horns and fill with cow dung and place the horn in the burial pit in October/November at 1" apart with the base downwards and surround them with 50% compost and soil. • Cover with soil and bury for 4 to 6 months and keep the burial pit moist and shaded at about 20^0 C temperatures free from weeds and earthworms. • After 4 months check for dung fermentation and if it is found dark they are ready for lifting.	Add 25 grams BD 500/acre in 15 litres rain/pure warm water (approx. 15-20 °C),stir for 1 hour alternately clockwise and anti-clockwise and spray in the late afternoon or evening (just before sunset), when Moon is descending. Spray 4 times a year – during the beginning and after rains, i.e. Feb-May-Nov-Dec. It can be stored in a glazed earthenware pots with loose fitting lids under moist and dark condition for one year.Result: Promotes root activitystimulates/ increases soil micro-life, regulates lime and nitrogen and helps germination and release of trace elements.
501: Cow horn-silica	• Crush silica quartz using a pounding rod, a mortar and pestle or hammer and grind to a fine powder. • Put 200-300 grams of powder quartz crystal per cow horn and moisten with water to make a stiff paste and fill the horn with such paste. • Bury horns in soil pit, 1 inch apart with base downwards, surround with 50% compost and soil from March/April to September	Apply 501two applications of BD 500. Apply during early morning 6-8 a.m. at sunrise during ascending Moon or Moon opposition Saturn. Prepare spray mixture by adding 1 gram silica in 15 litre of warm and quality water followed by thorough stirring.Spray the plants using a low-pressure sprayer (Knapsack 80-100 psi). Spray into the air to fall as a gentle mist over the plants. Spray twice during planting season. It can be stored for 3 years in a glass jar with a loose fitting lid. Helps photosynthesis and enhances light metabolism and improve colour, aroma and quality of plants.
BD 502 Yarrow (*Achillea millifolium*)	• Urinary bladder of the stag is used with Yarrow flowers in increase the cosmic effect. • Blow up the bladder with air when the bladder is fresh. Air dry and then collapse. At the time of use moisten to make it flexible.	• Use 1 gram each (502-506) for every 5 cubic metres of compost and 10 ml of 507 at 5% in 2-5 litres of water. These could be added to liquid manures and cow pat pits also.· Pits of 2ftx2ftx1-1.5 ft should be dug in a well fertile soil without trees nearby.

	• Cut the bladder, insert a funnel and introduce the flowers till the bladder is packed. Moisten the flowers with plant extract, stitch up the slit with cotton thread. • Store in a closed basket to keep away rodents/pests. Hang up in March to get cosmic influences. • Burry from September to March in mud pot with earth inside.	• Dig trench around pit to prevent weeds/roots. Mulch on top with coconut pith. Line the pit with bricks on the side but leave the bottom free • A marker should be clearly visible (e.g. brick lining) • Make a sign to define the preparation, date of burying and date of lifting, and a layout plan for record.
BD- 503 Chamomile (*Matricuria chamomilla*)	• Pick flowers of the Chamomile plant when petals are horizontal and keep in dry trays. • The intestine of a cow or bull should be cut into 15 cm bits and squeeze out undigested matter. Ties cut bit at one end with a cotton string & fill with dry flowers using a funnel. • Pack not too hard or loose. Stack the filled sausages into a bundle, which could be placed in a mud pot surrounded with fertile soil. Bury in October and let it remain in the soil till Feb/March.	• Maintain pit temperatures between 25-30^{0}C and maintain moistness by watering/sprinkling over the pits • Choose a moist, cool, dark location with good air circulation • Place the ready preparations into well labelled glass jars or glazed pots. • Place the pots into a well-insulated storage box using e.g. coir pith • Turn the preparations frequently, and maintain moisture
BD 504 Himalayan stinging nettle (*Urtica parviflora*	• Fill the dried leaves into terracotta pipes or mud pots and press well into the containers. Place the pot with lid under influence of Mars. • Moisten dry leaves with juice of leaves before filling if found dry. • Harvest leaves in May and September and lift the preparation in September after a year.	
BD-505 Himalayan oak bark (*Quercus glauca*)	• Crush the oak bark and place the crushed oak bark in the brain cavity of a skull of animal. Block the opening with a well-shaped bone piece and place the skull in a watery environment with weeds and plant muck which would have been damaged by the local diseases that affect the crop.	

	This helps build-up the resistance of the plants and follows the principles of Homeopathy. • It should be placed in a location where there is exchange of water such as rain drain/swamp. • Turn over frequently to correct the bad smell and formation of fungus. • The preparation is placed in September and lifted in March	
BD 506 Dandelion (*Taraxicum officinalis*)	• Place the dried flowers in the mesentery of a cow and wrap into a parcel and tie with a jute thread. • Place the parcel in a good mixture of soil and compost into a pot. • While lifting the preparation the mesentery may or may not be seen. • Place in September and lift in March.	
BD 507 Valerian (Valeriana officinalis)	• The juice of valerian flowers is prepared by grinding the flowers in a mortar and pestle. • This paste is added to water in the ratio of 1:4 in a bottle • Ensure storage in a cool place	
BD 508 (Equisetum arvense)	• Take 1 kg Equisetum arvense (Horsetail herb) or Casuarina and 10 litres water • Make a strong tea/tincture by boiling the Equisetum arvense or Casuarina in hot water for 2 hrs. Let it sit for 2 days.	Dilute the tincture: 50 grams tincture to 10 litres of water and spray onto the soil or over the plants in the early growing stages.For mild fungus problems BD 508 is often sufficient, but for more severe problems BD 501 is more effective.

(*Source:* Selvaraj, N., B.Anita, B.Anusha and M.Guru Saraswathi. 2006. Organic Horticulture creating a more sustainable farming. Horticultural Research Station, Udhagamandalam., TN Agriculture University Portal)

Fig. 3.6. Cow horn compost (BD: 500)

The BD preparations and their functions are given in Table 3.6.

Table 3.6. BD Preparations and their function

BD Prepn.	Herb or Material	Relationship to processes of	Planet	Function
502	Yarrow flower *Achillea millifolium*	Sulphur (S) Potassium (K) Trace Elements	Venus	Permits plants to attract trace elements in extremely dilute quantities for best nutrition
503	Chamomile flower *Matricuria chamomilla*	Calcium (Ca) Sulphur (S)	Mercury	Stabilizes Nitrogen (N) within the compost and increases soil life so as to stimulate plant growth
504	Stinging Nettle *Urtica parviflora*	Sulphur (S) Potassium (K) Calcium (Ca) Iron (Fe)	Mars	Stimulates soil health, by providing plants with the individual nutrition components needed, 'enlivens' the earth (soil).
505	Oak Bark *Quercus glauca*	Calcium (Ca)	Moon	Provides healing forces (or qualities) to combat harmful plant diseases.
506	Dandelion flower *Taraxicum officinalis*	Silicon (Si) or Silicic acid Potassium (K)	Jupiter	Stimulates relation between Si and K so that the Si can attract cosmic forces to the soil
507	Valerian flower *Valeriana officinalis*	Phosphorus (P)	Saturn	Stimulates compost so that Phosphorus component is properly used by the soil.

Source: TNAU Agricultural Portal, Organic Farming

These preparations are intended to help moderate and regulate biological processes as well as enhance and strengthen the life (etheric) forces on the farm. The preparations are used in very small quantity, i.e., a level teaspoon of BD preparation is added to 10 tons of compost.

Principles of Bio-dynamic farming

The BD preparations and BD compost may be considered the cornerstone of biodynamic agriculture. Both "biological" and "dynamic" qualities are complementary. These composts are used for managing soil health. The principles of bio-dynamic farming cover soil, organic matter, crop rotation, BD preparations, humus, cosmic force, farm organisms and substance, and energy.

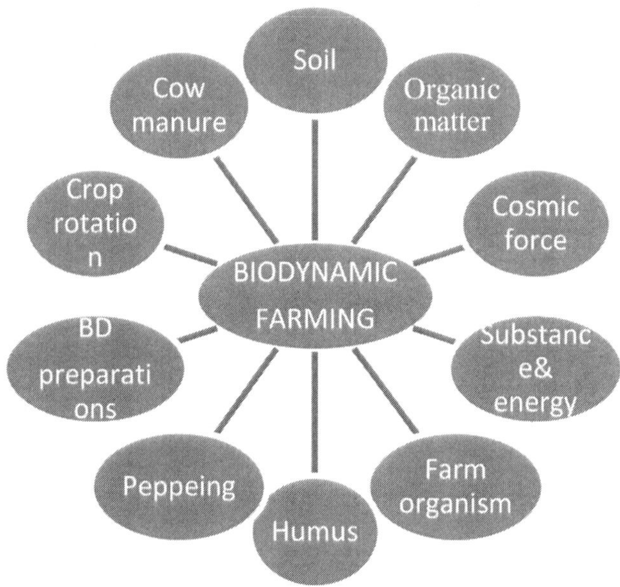

Fig. 3.7. Principles of bio-dynamic farming

Substance and energy: Life depends on the interaction of matter and energies. The interaction of substance and energy forms a balanced system. Only plants which have grown in a balanced soil can give us energy as well as substance.

Soil: The soil is a living system which contains nutrients, micro-organisms, worms and other living bodies. If the soil is balanced in its life forces, the plants growing in it will be stronger, healthier, and higher in quality.

Organic matter & humus: Soil organic matter is enhanced by building compost heaps and using the biodynamic compost preparations. Humus is completely digested crude organic matter which holds the fertility of the soil in a stable way and retains water. It is the base for building up the soil and fostering its formation is the first priority when converting to biodynamic farming.

Cow manure: Cow dung is special manure because of the lengthy digestion process of the cow which adds much beneficial bacteria to the substance. It is used in building the compost heaps as a starter for the biodynamic preparations.

Cosmic forces: The influences of heavenly bodies on plant growth are recognized by using the preparations and following the sowing calendar.

Biodynamic preparations: These preparations are used to enhance the soil and the plants, and breaking-down process and potential life forces in the compost heaps.

Crop rotation: Crop rotation improves soil structure, water and temperature balance in the soil and controls weeds and pests.

Peppering: One may collect the insects, weed seeds, or dead animal skins, burn them at the appropriate time according to planetary positions and the ash may be mixed with water and spray it on the land as an alternative to chemical sprays.

The farm organism: Important aspects of the farm include the water source and balance, prevention of soil erosion by planting trees and hedges which also help in wind protection and providing animal habitat.

Weeds, pests and diseases

Weeds growing in specific places show a deficiency in the soil, as pests and diseases show a shortcoming in agricultural practices. It is well known and proven that insect pests and diseases will only attack weak plants.

3.4.2.5 Regenerative agriculture

Among various alternative agriculture practices, regenerative agriculture is a farming system that focuses on rehabilitation and improvement of the ecosystem though soil health improvement, water management and input use. The primary objective of regenerative agriculture is to regenerate soils and increase their organic content to improve their fertility. It conserves and restores soil organic matter by protecting the habitats of micro and macro-organisms in the soil. Regenerative agriculture practices are based on three key principles that aim to reduce tillage, maintain a better balance, and continuously nourish the soil with permanent plant cover.

One goal of regenerative practices is to use some of the carbon that plants have absorbed from the atmosphere to help restore soil carbon. Regenerative agriculture focuses on top-soil regeneration by organic recycling, increasing biodiversity and improving water cycle. It is based on philosophy of agro-ecology, permaculture, agroforestry etc. for a holistic management that is climate resilient. Practices for regenerative agriculture include no-till agriculture or reduced tillage and organic waste recycling mainly for topsoil regeneration and increasing biodiversity. It provides resilience to climate change and improves soil health. When soil health improves, the input requirement is reduced and the plant may be resistant against weather aberration and pest attack. Regenerative agriculture captures atmospheric carbon dioxide by growing plants that move that carbon dioxide into the soil.

Principles and practices

Regenerative agriculture is based on principles of increasing soil fertility and improving whole agro-ecosystems (soil, water, and biodiversity). The practices like permaculture, aquaculture, agro-ecology, agroforestry and livestock farming are usually adopted in this type of agriculture. Conservation farming, no-till farming, minimum tillage, and pasture cropping, cover crops & multi-species cover crops, organic annual cropping and crop rotations are various techniques in such farming system. Natural sequence farming, grass-fed livestock, polyculture and full-time succession planting of multiple and inter-crop sowing or plantings, borders planting for pollinator habitat and other beneficial insects, agro forestry, silvi-pastoral systems are some of the options for regeneration agriculture. Regenerative agriculture is a holistic system which takes care of soil ecosystem, natural cycles and a sustainable production system in partnership with nature.

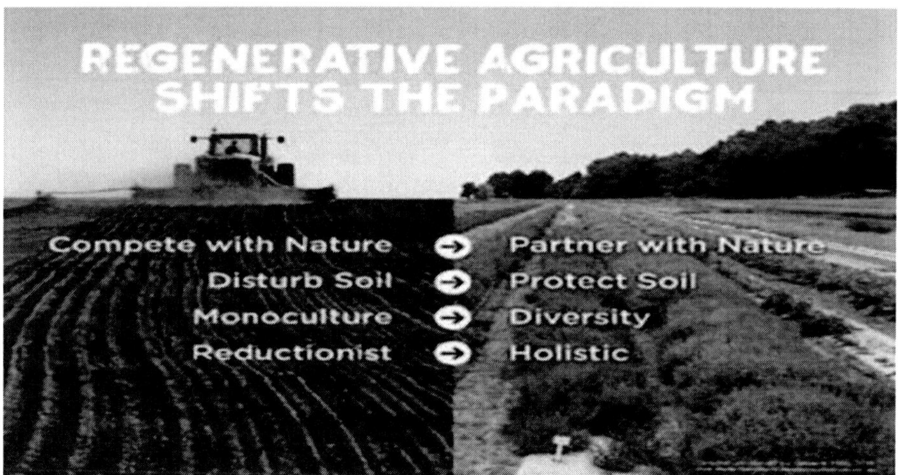

Fig. 3.8. Shift to regenerative agriculture

Key techniques

- **Conservation tillage**: Tillage creates soil compaction, causes erosion and releases carbon dioxide to the atmosphere. No-tillage or minimum-tillage minimises soil disturbance, increases soil organic matter, store organic carbon in soil and makes the plants resilient to climate.

- **Diversity:** Release of carbohydrates by plant roots differ by the plant type and soil microbes engaged in in nutrient cycle. Increasing plant diversity the farmers can create nutrient-dense soils which are more productive.
- **Cover crops and rotation:** Bare and uncovered soil loses nutrients by erosion for which cover cropping builds the nutrient level in the soil. Crop rotation with inclusion of legumes in the rotation will increase organic matter content of soil, increase nitrogen through fixation by legumes, and keeps the pest attack in check.

Fig 3.9. Cover crop and mulching

Fig. 3.10. Crop rotation

Common practices under regenerative agriculture

The common ten practices under regenerative agriculture are: (i) minimum-tillage/zero tillage, (ii) cover cropping, (iii) composting, (iv) increasing crop diversity, (v) soil microbe food, (vi) annual organic cropping, (vii) integration with livestock, (ix) silvi-pasture, and (x) agro-forestry.

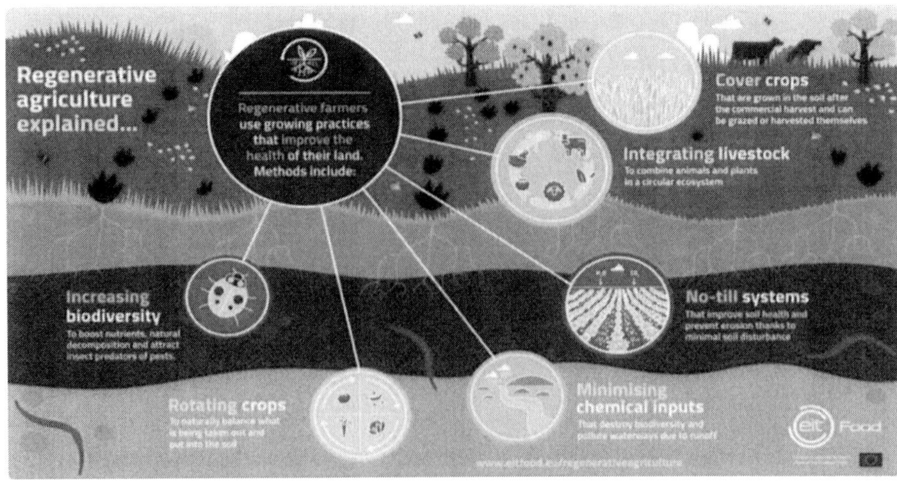

Fig. 3.11. Regenerative agriculture practices.

Benefits

Regenerative agriculture makes the soil more resilient to impact of climate change like flood and drought by increasing soil biodiversity and soil health. The other benefits are reduction of soil erosion, production of nutrient dense crops, water retention in soil and carbon sequestration in soil.

3.4.2.6 Homa farming

Homa farming is a form of organic farming which is believed to be the revised version of vriksha ayurveda (ayurveda for the plant kingdom). Agnihotra is the basic technique practiced in Homa organic farming where dried cow dung, ghee of local cow and unpolished brown rice are burned in an inverted pyramid shaped copper vessel along with singing of special mantras. It is believed that the mantras sung in resonance with a bio-rhythm activate special vibrations which produce a particular healing atmosphere (Shinogi, KC et al, 2016).

Homa preparations like Agnihotra ash powder and Agnihotra ash solution may be used sometimes to control serious pest and disease attack. Another famous Homa organic farming input is Gloria Biosol, a liquid biofertilizer based on Agnihotra ash prepared under anaerobic conditions in a bio-digester. Biosol was developed by Gloria Guzman Mendez in Peru, South America. It was prepared out of vermi-compost, fresh cow dung, cow urine, Agnihotra ash, copper shree yantra disc (a source of energy attractor) and water. This preparation is mainly used to rejuvenate the plants and also to enhance plant growth by controlling pest and diseases.

Research conducted in India showed that fumes emanating from Agnihotra eradicate microorganisms that cause illness and diseases (Pachori et al, 2013). Burning of cow dung for disinfection is an age old practice. According to Russian scientist Sirovish, cow's ghee also has immense power to protect human body from the ill effect of radioactive waves. It is also reported that when cow's ghee is burned with rice it produces gases like ethylene oxide, propylene oxide, formaldehyde, and beta-propiolactone which have inhibitory effect on microorganisms (Monskar, 1982; Potdat, 1992).

Agnihotra ash is beneficial at all stages of farming operations like soil treatment, water treatment, seed treatment. However, the basic intercultural operations and composting are also necessary along with the application of Agnihotra ash.

There are a number of research reports about the beneficial effect of Agnihotra ash. Kratz and Schnug (2007) reported that Agnihotra ash improved the solubility of phosphorus in soil. The practitioners and propagators of homa farming call it a "revealed science". It is an entirely spiritual practice that dates from the Vedic period. The basic aspect of homa farming is the chanting of Sanskrit mantras (agnihotra puja) at specific times in the day before a holy fire. The timing is extremely important. Homa Organic Farming is a system of agriculture that may be added to any good organic farming practices.

Treating the atmosphere

The main difference between Homa organic farming and other organic farming techniques is that Homa farming regards the atmosphere as the most important source of nutrition, whereas in any other farming practices today the atmosphere is almost totally neglected. Ancient science of Homa Therapy states that more than 75% of nutrition to plants and soil comes through the atmosphere.

Homa organic farming injects nutrients into the atmosphere to nourish the plants, to prevent disease, bring natural predators and nutritious, timely rains. Homas are specially prepared fires for the purification of the atmosphere. If you make the atmosphere more nutritious and fragrant by preforming Homas, a type of protective coating comes on plants and diseases, fungi, pests, etc. do not thrive. Plants' capacity to breathe increases and the toxic choking effect due to atmospheric pollution is eliminated.

AGNIHOTRA is the basis of HOMA Farming. This should be performed twice daily in the Agnihotra hut. If cow dung and ghee are readily available they should be used in greater quantity when doing Agnihotra. The Agnihotra pyramid is a generator of life-sustaining energies.

Om Tryambakam Homa should be done as many hours as possible, up to four hours daily and up to 24 hours on full moon and new moon days. Om Tryambakam Homa enhances the healing energy cycle set up by daily Agnihotra fires, injecting more nourishment into the atmosphere. *Vyahruti Homa* can be practiced at any time, except Agnihotra time. It is also performed at the beginning of Om Tryambakam Homa. In order to start Homa Organic Farming the farmer will need to establish a Resonance Point on his farm.

Resonance points

Resonance Technique is a part of HOMA Organic Farming, where simple practices are used to heal large areas of diseased land in a short time. One Resonance Point can heal up to 200 acres (80 hectares) of land. The same human effort is required to heal one or two hundred acres. For this, 10 new copper pyramids are needed which are activated with Mantra and placed on the farm in a special configuration by a Homa Therapy volunteer who is authorized to install resonance points.

Also two simple huts have to be built with inexpensive, natural materials found locally, such as wood, adobe bricks, mats, bamboo, stone, cane, etc. Nobody will live in these huts. They are simply to protect the person performing the HOMA healing fires from the sun and rain and to prevent animals such as dogs, cats, chickens, etc. from entering.

Procedure of Agnihotra preparation (Jani C. P.2020)

A few minutes before the actual time of sunrise and sunset, start to prepare the Agnihotra fire as follows:

1. Place a flat piece of dried cow dung at the bottom of the copper pyramid. Arrange pieces of dried cow dung in the pyramid in such a manner as will allow air to pass
2. Apply a little ghee to a small piece of cow dung and light it. Insert this lighted piece of cow dung in the middle of the pyramid. Soon all the dung in the pyramid will catch fire. However, do not blow on the fire so as to avoid bacteria from the mouth affecting the fire.
3. Take a few grains of rice in a dish or left palm and apply a few drops of ghee to them.
4. Exactly at sunrise utter the first Mantra and after the word SWAHA add a few grains of rice (as little as you can hold in the pinch of fingers will sufficient) to the fire. Utter the second Mantra and after the word SWAHA add a few grains of rice to the fire. This completes morning Agnihotra.
5. At sunset do the same by using evening Mantras. This completes evening Agnihotra. After each Agnihotra try to spare as many minutes as you can for meditation. You can sit at least till the fire gets extinguished itself.
6. Just before the next Agnihotra collect the ash and keep it in a glass or earthen container. This highly energized ash can successfully be used as organic fertilizer.

Application of Agnihotra

Agnihotra can be used for seed treatment to increase the seed germination. Mixture of cow urine and water in a ratio of 50:50, 4 tablespoons of Agnihotra ash per 5 litres of solution are added and stirred. Seeds and seedlings should be soaked in this solution for 30-40 minutes. Like cow dung, cow urine has antibacterial effects and provides a protective coating around the seeds and seedlings. They should be dry enough to spread, but moist enough so that the core of the seed doesn't dry out. Seedlings may be planted immediately after being treated with the solution. Agnihotra ash is applied as fertiliser as it contains

97% P, 2.32% K and 0.34% N. It can be dusted after irrigation gives nutrition as natural fertilizer.

Treating the seeds

The seeds should be organic and not treated with chemicals.The seeds are placed in the Resonance Hut on the eastern side of Agnihotra pyramid and exposed for two minimum Agnihotra fires. In Homa Therapy seeds are treated in the following method.

1. Place the seeds in cups or jars and label each container with the name of the seed.
2. Cover the seeds with cow's urine and soak for 1-2 hours depending on the size of the seed.
3. Drain the seeds and cover them in moist cow dung. The mixture can be spread on plates for semi-drying. The treated seeds are then sown in lines. If the seeds are not too small , they may be wrapped in an individual bit of cow dung.

For larger applications/broad acre farming

Cover seeds with cow's urine (diluted 1:10 if necessary) and soak the seeds for one to two hours, depending on the size and nature of the seed. After soaking discard the remaining cow's urine and cover seeds with fresh cow dung slurry and Agnihotra ash; best to do this in a bucket so that all the seeds are covered. After drying a little, the seeds can be sown and covered with soil. If planting seed with a seed drill, mix seeds with cow dung slurry and Agnihotra ash, let the seeds dry. When using the seed in the seed drill, the seed may not be mixed with cow dung. In case treating with cow urine makes the seed too soft, treat the seeds with Agnihotra ash and sprinkle a bit of cow urine just before planting.

Agnihotra ash – a powerful tool

Agnihotra ash is the Secret Weapon of the Homa Organic Farmer. Agnihotra ash is extremely medicinal.

Agnihotra ash powder can be applied to

- Well water and other irrigation sources, such as tanks or lakes
- Soil before ploughing – just before ploughing sprinkle ash on the land (not advisable in windy conditions)
- Furrows while planting the seeds

- Seeds before sowing
- Roots while transplanting
- Plant leaves to protect against insect and fungus attack
- Soil around the plants

To make Agnihotra ash water solution

1. Put about 250g Agnihotra ash into 200 litres of clean water.
2. Stir it.
3. Leave it in the sun for 3 days. In case of rain it should be covered.
4. After 3 days it is ready to be used.
5. Strain through a cloth.

Agnihotra ash water solution is very useful for

- Natural control of difficult pests
- Encouraging rejuvenation
- Enhancing plant growth

Agnihotra ash water solution can be applied

- Directly on the soil
- Around individual plants
- To irrigation water
- As foliar spray
- As a good alternative to Agnihotra ash powder when its direct use is difficult (e.g. in strong winds)

Tree paste

Tree paste can be made with Agnihotra ash, mud and water, best Agnihotra ash water. The mixture can be applied on wounds of the tree.

Biosol

Homa biosol is a liquid bio-fertilizer based on Agnihotra ash which was developed by Gloria Guzman Mendez in Peru, South America. It is prepared under anaerobic conditions in a bio-digester. Biosol can be used as a foliar application to nourish plant kingdom or applied directly to the soil to rebuild soil health.

Preparation of Biosol (for 500 litre tank)

Vermicompost	:	80 kg
Fresh cow dung	:	80 kg
Cow urine	:	10 lit
Agnihotra ash	:	250 gm
Shree yantra	:	1 unit
Water	:	200 lit.

Method of preparation

1. Put one copper Shree Yantra at the bottom of the tank facing upwards. (Shree Yantra geometrical design is engraved in copper and, according to traditional knowledge, is a powerful energy attractor).
2. Collect 80 kg fresh cow dung, 80kg vermicompost and 10 litres cow urine.
3. Divide the cow dung and vermicompost into 3 piles each.
4. Mix 1 part vermicompost, 1-part cow dung, about 3 litres of cow urine and about 50 litres of Agnihotra ash water solution and stir to a slurry and pour it into the tank.
5. Repeat the same process with the second portion of the materials while stirring the material continuously and then add the second slurry to the tank.
6. Finally repeat the process a third time and add the slurry to the tank.
7. Add the remaining Agnihotra ash water solution to the tank and again stir

Application of gloria biosol

1. It should be used for foliar applications with water at a ratio of 1:15 to 1:20 depending on density of plant population.
2. We can spray Biosol liquid on any type of crop at an interval of seven days. Roughly 20 litres is required per acre per month.
3. If we preserve Biosol liquid in airtight cans it will last longer, say about six months.
4. Left over solid Biosol which is having maximum macro nutrients should be mixed with any type of organic manure at a ratio of 1:5.

3.4.2.7 Low input sustainable agriculture (LISA)

Low-Input Sustainable Agriculture (LISA) is an ecological, organic, regenerative, biological, or simply alternative agriculture. The main goals of LISA are profitable and productive farming, and protection of resources and environmental quality. It ensures safe and nutritious food supplies.

Techniques

LISA techniques include rotations, crop and livestock diversification, soil and water conserving practices, mechanical cultivation, and biological pest controls. "Low-input" is a catchword for what many feel is a primary requirement for economic and environmental sustainability in farming-the need to cut back on purchased off-farm inputs. These especially include synthetic chemical fertilizers and pesticides, but also livestock growth stimulants. Crop rotation, soil building practices and diversification are important tools.

Legume rotations and use of green manure (crops planted specifically to be ploughed under to enrich the soil) can supply plant nutrients. Soil and water conserving practices, including or in combination with rotations, enhance soil quality and productivity. Crop rotations also help control weeds, insects, and plant diseases. With LISA, pests can be prevented or controlled without using chemicals. Mechanical cultivation can substitute for chemical weed killers. Weeds can be controlled by rotation. For example, rotations and crop diversification may include a crop like rye specifically because it is toxic to weeds. Integrated pest management can play an important role also. Scouting of fields to monitor insect infestations is one way to limit the use of insecticides to a when-needed basis. Biological techniques, such as use of beneficial insect predators, can often eliminate the need for insecticides entirely.

Fig. 3.12. Low input sustainable agriculture (LISA)

Summary

Alternative agriculture practices are production systems that do not use conventional methods and works with nature. In spite of different types of systems such as organic farming, permaculture, regenerative agriculture, natural farming, bio-dynamic farming, low-input sustainable agriculture etc., they follow the concept of agro-ecology and are designated to drastically reduce or eliminate the inorganic fertilisers and chemical pesticides that are mostly used in conventional agriculture. Despite their differences, these systems share common values. They were firstly thought as ways to preserve the environment and more precisely soil and water and make the farming resilient to weather aberrations. They intend to comply with natural cycles, by using crop rotations, cover crops or no-tillage, etc. Farmers who practice alternative agriculture are able to ensure a profit and respect the environment and people.

References

Albert Howard (1873-1947), An Agriculture Testament

Ammann, K. (2008). Integrated farming: why organic farmers should use transgenic crops. New Biotechnology, 25(2/3), 101-105.

Anderson, J. 1985. Farm programs and alternative agriculture. In Proposed 1985 Farm Bill Changes: Taking the Bias Out of Farm Policy. Proceedings of the Institute for Alternative Agriculture Second Annual Scientific Symposium. Institute for Alternative Agriculture, Greenbelt, Maryland.Google Scholar

Aramo. 2012 ,"Rainwater harvesting". Germany:. Archived from the original on 6 June 2013. Retrieved 19 August 2015.

Bell, G. (2004). The permaculture way: Practical steps to create a self-sustaining world. Hampshire, United Kingdom: Permanent Publications.

Carson, R. (1962) Silent spring. 1st Edition. Boston: Houghton Mifflin

Chalker-Scott, L. (2004) The myth of biodynamic agriculture: Horticultural myths. Washington, DC: Washington State University.

Code, J. (2014) Muck and mind: Encountering biodynamic agriculture. Great Barrington, MA: Lindisfarne Press.

Fukuoka, M. (1978) The one straw revolution. Emmaus, PA: Rodale Books

Gillman, P. (2008) The truth about organic gardening: Benefits, drawbacks and the bottom line. Portland, OR: Timber Press.

Harwood, R. 1984. Organic farming research at the Rodale Research Centre. In Bezdicek, D., Power, J. F., Keeney, D. R., and Wright, M. J. (eds.). Organic Farming: Current Technology and Its Role in a Sustainable Agriculture. Agronomy Society of America, Crop Science Society of America, and Soil Science Society of America. Madison, Wisconsin, pp. 2–3.Google Scholar

Holger, K. (1994) Biological dynamic farming – an occult form of alternative agriculture? Journal of Agriculture and Environmental Ethics, 7(2), 173–187.

Holmgren, D (2013), "Permaculture: Principles and Pathways Beyond Sustainability".

Holmgren, D. (2003). Permaculture: Principles and pathways beyond sustainability. Holmgren Design Services.

Holmgren, D. (2000) Permaculture. Holmgren Design Services. http://www.holmgren.com.au/contact

Holmgren, David (2002). Permaculture: Principles & Pathways Beyond Sustainability. Holmgren Design Services

Holzer, S. (2004). Sepp holzer's permaculture: A practical guide to small-scale, integrative farming and gardening. White River Junction, Vermont: Chelsea Green Publishing Company

Hurter, U. (ed.) (2014) Agriculture for the future: Biodynamic agriculture today. Dornach: Verlag am Goetheanum.

IFOAM (2007). Final draft on the definition of organic agriculture. http://www.ifoam.org/growing_organic/definitions/sdhw/pdf

Koepf, H.H. (1989) The biodynamic farm: Developing a holistic organism. Great Barrington, MA: Steiner Books.

Lampkin, N.H (1994),Organic farming: Sustainable Agriculture in practices-pp1-9, In: The economics of organic farming-and international perspective (Lampkin, N.H, and Padel,S.eds.). CAB International, UK

Lee, D. R. (2005). Agricultural sustainability and technology adoption: issues and policies for developing countries. American Journal of Agricultural Economics, (87)5, 1325-1334.

Lillington, Ian; Holmgren, David; Francis, Robyn; Rosenfeldt, Robyn. "The Permaculture Story: From 'Rugged Individuals' to a Million Member Movement" (PDF). Pip Magazine. Retrieved 9 July 2015.

Lorand, A (1996) Biodynamic agriculture: A paradigmatic analysis. http://www.triptolemos.net/wp-content/ uploads/2015/07/Lorand-short-Dissertation.pdf.

Masson, P. (2012) A biodynamic manual: Practical instructions for farmers and gardeners. Edinburgh: Floris Books.

Mollison, B. (1988) Permaculture: A designer's manual. Tasmania: Tagari.

Mollison, B. C. (1991). Introduction to permaculture. Slay, Reny Mia., Jeeves, Andrew. Tyalgum, Australia: Tagari Publications. ISBN 0-908228-05-8. OCLC 24484204

Mollison, Bill (15–21 September 1978). "The One-Straw Revolution by Masanobu Fukuoka". Nation Review. p. 18.

Mollison, B., & Holmgren, D. (1979) Permaculture two. Tasmania: Tagari.

Pfeiffer, E. (1947/2004) Soil fertility, renewal and preservation: Biodynamic farming and gardening. 2nd edition. Sussex, UK: Lanthorn Press.

Selvaraj, N., B.Anita, B.Anusha and M.Guru Saraswathi. 2006. Organic Horticulture creating a more sustainable farming. Horticultural Research Station, Udhagamandalam

Steiner, R. (1983) The boundaries of natural science. New York: Anthroposophic Press

Steiner, R. (1993) Agriculture. Pennsylvania, PA: Biodynamic Farming and Gardening Association Inc. Steiner, R. (2004) Agriculture course: The Birth of the Biodynamic Method. East Sussex, UK: Rudolf Steiner Press.

Thornton Smith, R. (2009) Cosmos, Earth and nutrition: The biodynamic approach to agriculture. Forest Row: Sophia Books.Organics

TNAU Agricultural : Organic Farming

Willer, H & Lernoud, J (2019): The World of Organic Agriculture Statistics and Emerging Trends 2019

Woese, K., Lange, D., Boess, C., & Bogl, K.W. (1997) A comparison of organically and conventionally grown foods: Results of a review of the relevant literature. Journal of the Science of Food and Agriculture, 74. 281–293.Permaculture.

4

Soil and Plant Nutrition

4.1 Soil and its characteristics

Soil is made up of minerals, organic matter, gases, liquids and micro-organisms in different proportions. It provides anchorage to plants. It is important for life as it provides the medium for the growth of a plant, habitat for several insects and other organisms. Soil is a store house of nutrients, micro-organisms, air and water. It is the source of four essential 'living' factors including food, clothes, shelter, and medicines.

Soil minerals are divided into three size classes such as clay, silt, and sand. The percentage of these particles in these size classes is called soil texture. The arrangement of these particles is called soil structure. Soil colours range from the common browns, yellows, reds, greys, whites, and blacks to rare soil colours such as greens and blues.

Soil horizon refers to different layers of soil which interact with each other. Although there is complexity and diversity in such horizons, the surface horizons are rich in life and organic matter. The horizons below the surface horizons are formed through diverse soil formation processes. Below the surface horizons we see ultimately unaltered layers of parent material.

These master horizons may then be further annotated to give additional information about the horizon. Horizons are first assigned to one of the following master horizons as designated by a single capital letter:

Name	Layer	Characteristics
O	Humus or organic	Horizon 'O' is very thin or thick in different soils and may not be present at all in others. It contains high percentage of soil organic matter due to decomposition of leaves.
A	Top-soil	Looks dark by accumulation of organic matters. Minerals from parent material with organic matter are incorporated. Very suitable for plants and other organisms to live.
E	Eluviated	Horizon formed through the removal (eluviation) of clays, organic matter, iron, or aluminium and usually lightened in colour due to these removals. Missing in some soils but often found in older soils and forest soils

B	Sub-soil	Broad class used for subsurface horizons that have been transformed substantially by a soil formation process such as colour and structure development or intense weathering leading to the accumulation of weathering-resistant minerals. Rich in minerals that leached (moved down) from the A or E horizons and accumulated here
C	Parent material	A horizon minimally affected or unaffected by the soil formation processes. The deposit at Earth's surface from which the soil developed.
R	Bed rock	A mass of rock such as granite, basalt, quartzite, limestone or sandstone that forms the parent material for some soils – if the bedrock is close enough to the surface to weather. This is not soil and is located under the C horizon

Fig. 4.1. Soil horizons

4.2 Functions of soil

Food, water and energy security are required for sustenance of humankind and protection of the environment. Soil is considered as a critical component of food, energy and water security. Soil is our life support system that provides anchorage for roots, hold water and nutrients. They are home to innumerable microorganisms that fix atmospheric nitrogen and decompose organic matter, and silent armies of microscopic animals as well as earthworms and termites. Soil is a reservoir for nutrients and water required for the plant growth and development. It also provides an environment for the break-down and immobilisation of materials added to the soil surface. Healthy soil gives us clean air and water, products from agriculture and forestry and rangelands. The soil performs many functions which include functions related to natural ecosystems, agricultural productivity, maintenance of environmental quality etc.

i) **Soil ecosystem:** Soil is formed by weathering of rocks which is a complex mixture of minerals, water, air, and organisms. Pedogenesis involves interactions between the lithosphere, hydrosphere, atmosphere, and biosphere (Bosch, C ,1988).The spatial arrangement of the voids, the pore system, and the solid soil components determines the soil structure, which may be primary and secondary. The soil solid phase can be separated into various particle sizes by classification processes such as sieving and sedimentation. The total organic fraction of the solid phase of the soil is divided into biomass, which includes all living organisms (soil flora and fauna) and dead organisms and their transformation products. Within their biotope, the soil organisms form a community known as an edaphon which is divided into micro flora (bacteria, fungi, algae), micro fauna (protozoa), mesofauna (nematoda, rotifera, tardigradae, mites, and collembola), macrofauna (annelids, snails, insects and their larvae), and megafauna (vertebrates). Almost 100 % of the elemental composition is accounted for by C, O, H, N, and S. The central importance of the soil is due to the fact that fundamental ecological processes interact there. Hence the soil is an essential basis for all life processes with interactions of the biological, chemical, and physical environments that result in the transformation of such materials. This immobilization may enable breakdown of the potentially dangerous materials to less dangerous forms.

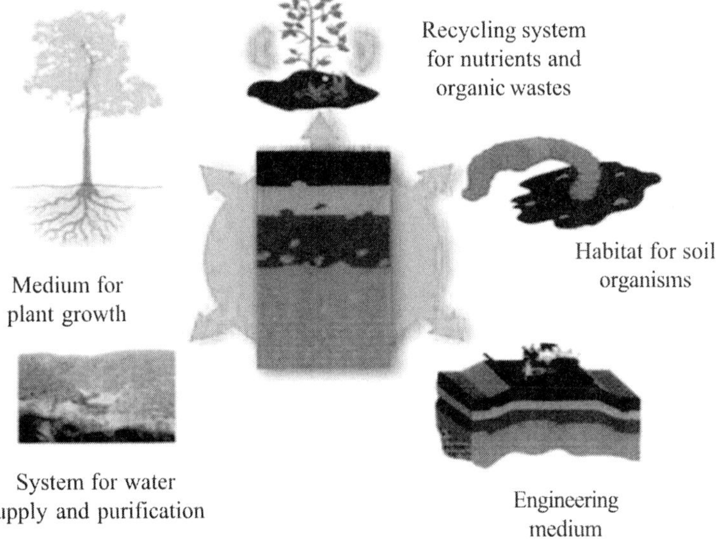

Fig 4.2. Ecological services of soil

ii) **Ecological function**: The main ecological functions of soil include nutrient cycling, carbon storage and turnover, water maintenance, soil structure arrangement,regulation of aboveground diversity, biotic regulation, buffering, and the transformation of potentially harmful elements and compounds , e.g., heavy metals and pesticides (Haygarth and Ritz, 2009). The dominant group of soil biota are microorganisms comprising bacteria, archae, and fungi. The rhizosphere contains nematodes, micro-arthropods, enchytraeids and earthworms. The soil surface and litter layer also contains large numbers of micro fauna species such as beetles, spiders, diplopods, chilopods, snails etc. Activities of soil biota regulate the ecological functions of soil and these activities are considered as indicators of soil quality.

The regulating function of the soil depends on the regulation of natural material and energy cycles. Soil is a storehouse of water received from hydrological cycle. It can regulate the drainage, flow and storage of water and solutes, which includes nitrogen, phosphorus, pesticides, and other nutrients and compounds dissolved in the water. Soil water is available for crops within the limits of field capacity and permanent wilting coefficient.

Filtration as defined as mechanical retention of material >100 nm in size (Blume, HP, 1990) occurs during the flow of aqueous suspension, through the pore systems. Infiltration of soil water is one of the most important effects counteracting erosion. Soil compaction may reduce rate of infiltration. Pore blockage can also have deleterious effects on the transformation and habitat functions (stagnant water, air deficiency). The buffer function encompasses the buffering of water balance, buffering of acid inputs and buffering of nutrients and pollutants. The buffering of water balance is largely a result of the ability of soils to store water in pore spaces by means of capillary forces. The water-storage capacity of a soil is characterized by the amount of water that can be retained for a given value of capillary pressure (pF characteristic). Buffering of acid inputs is achieved by various soil buffer systems, depending on the pH range. Buffer systems consist of a weak acid or base and the corresponding salts, and maintain constant pH within a certain range. All soils are exposed to acids which are added from natural sources (absorption of nutrient ions) and human factors (addition of fertilisers etc. The buffering of nutrients and pollutants possible by presence of organic and inorganic exchangers, affects most ecological soil functions. For nutrients and inorganic pollutants, ion formation is the most important process. Ion exchangers have positive or negative excess charge, thus binding anions or cations. Sorption of this type is reversible and an equilibrium is established in the soil solution. The input of nutrients and inorganic pollutants into the soil, their dissolution in soil water, buffering by adsorption and desorption, uptake by plants, and loss of eluates can be depicted in nutrient and pollutant cycles.

iii) **Bio-physical functions**: These functions of agricultural importance include nutrient cycling, water dynamics, filtering and buffering, anchorage to plants and promotion of biodiversity and habitat. Soil supports the growth of a variety of plants, animals, and soil microorganisms, usually by providing a diverse physical, chemical, and biological habitat.

The organic material of the soil consists of 85% post-mortal substance, 10% plant roots, 4% bacteria and fungi, and 1% soil fauna. Bacteria and fungi account for 90% of degradation, while soil animals are responsible for only 10%. However, soil animals are highly effective in accelerating microbial degradation and can increase the availability of nutrients by up to 50%. Field studies have shown that decreasing their numbers decreases the degradation performance of the soil by up to 40%. In general, single-cell organisms, worms, and arthropods each make up one-third of the soil biomass.

Respiration of an organism is a measure of its metabolic activity and, therefore, its material turnover. Protozoa account for 70% of respiration–considerably higher than their biomass fraction of about one-third. It is not general knowledge that these soil-inhabiting micro-fauna produce approximately as much biomass as earthworms. In the food chain, protozoa represent an ecologically important link between bacteria and small metazoans (e.g., rotifera and nematode) and thus exhibit pronounced catalytic activity towards bacterial processes and re-mineralization processes in the soil. The occurrence of protozoa is mainly limited to the upper 10 cm of the soil. Due to their small size or flat, flexible form, flagellates and naked amoebas can enter even the smallest soil pores.

iv) **Productive function:** Soil has the ability to maintain its porous structure to allow passage of air and water, withstand erosive forces, and provide a medium for plant roots. The productive function is the ability of the soil to act as substrate for cultivated plants (food and feed plants, renewable resources). Effects on the productive function are generally assessed in economic terms, rather than ecologically, since cultivation in agriculture, forestry, and horticulture is intended to produce a profit.

Soil stores and supplies plant nutrients and other elements for growth of plants through nutrient cycles. During these biogeochemical processes, nutrients can be transformed into plant available forms. A part of such nutrients may also be lost to soil and atmosphere. Plants obtain not only their nutrient elements, but also pollutants, mainly from the soil. In modern agriculture inorganic and organic crop-protection agents, organic and mineral fertilizers, and organic and mineral wastes are applied for higher yields, all of which can lead to soil pollution and degradation. The behaviour of crop protection agents (CPA) in soil depends on their properties, the way in which they are used, and on local conditions. The amount of CPA applied, the frequency of application, and the time at which it is applied are relevant to movement of the agents in the soil and their potential entry into water systems. The manner in which a CPA is used affects its degradation, which is faster at the higher temperatures of spring and summer and leads to a more rapid decrease in the amount of CPA present.

Fertilization and the utilization of waste materials can lead to the introduction of both organic and inorganic pollutants into the soil. In contrast to the application of crop-protection agents, this is an unintentional contamination. The aim of fertilization, which can include utilization of waste, is to improve conditions for the growth of cultivated plants, either by direct action on the plants themselves or by affecting the soil. The macronutrients like nitrogen, phosphorus, and potassium are the most important for the nutrition of cultivated plants. Thus in the sustainable development of agriculture, both nutrient and pollutant aspects must be taken into account. According to Klapp (Klapp, E. 1958) the most important aspect is the sustainability: the long-term ability of the soil to provide all requirements for plants to flourish, that is, to provide constant or only slowly decreasing harvests without fertilization. In contrast, in modern agriculture, fertilization is regarded as a definite means of supplying required plant nutrients to soil. The starting point for sustainable use of the soil should be a nutrient supply that is appropriate to the needs of the plants and which avoids an excess in the soil and release into other environmental compartments.

Residual material remaining on the field after harvesting must also be included in the balance, and manure should be used in preference to external waste materials. Mineral fertilizers, organic fertilizers, and wastes all introduce inorganic and organic pollutants into the soil and can cause adverse effects there and in other environmental compartments. Mineral fertilizers can contain various amounts of pollutants, depending on the origin of their raw materials and the production process used. The main pollutants entering soils via fertilization are Cd and Cr, and to a lesser extent Pb, Ni, and As. Most of the heavy metal load comes from phosphate rock and P fertilizers derived therefrom (super and triple phosphate, partially digested and soft phosphate rock). The specific quantity of heavy metals introduced is strongly influenced by the various land-use and fertilization systems. In terms of quantity, the most important organic fertilizers are liquid manure, sewage sludge, and compost. For compost, numerous quality marks exist, whereby the pollutant concentration is specified. If an average application of 10 t per ha is assumed, a cadmium load of 10 – 40 g/ha can be calculated (Bannick, C and Bodenk, M, 1944) . However, depending on the region and the spectrum of discharge or compost starting materials, these theoretical pollutant loads are not always fully exploited.

In forestry, atmospheric nutrients and pollutants are of major importance. The problem of pollution in forestry is exemplified by a comparison between Cd inputs and outputs.The mobilization of cadmium accumulated in the forest soil as a result of acidification demonstrates the problems that arise when the filter capacity of the soil is exhausted. The filtration capability of the soil can only be regarded as undisturbed if the output (at the background level) is lower than the current input. If changes in soil properties lead to release of accumulated materials, then pollution of the groundwater or of the habitat for organisms cannot be excluded

v) **Other Functions:** The exploitation of waste is possible in cases where it is both safe and useful. Utilization of waste should be safe with regard to nutrients and pollutants. Soils should be included under mineral wastes because the organic matter (up to 30 %) they contain differs strongly from organic wastes in type and constitution. In the utilization of organic wastes, the pollutant load is the most important factor, since they remain almost entirely on the surface after mineralization of the organic component, contributing to an increase in the pollutant potential. In the case of mineral wastes, there is generally no loss of mass after use in the soil, so that a different pollution assessment concept must be used. The pollutant concentrations in the waste should not exceed the background concentrations in the location of use.

4.3 Soil properties

Soil is composed of mineral matters, air, water, and different living and non-living organic compounds. In general, the abiotic component of the soil accounts for about 40-45% of the soil volume followed by air and water that occupy 25% each with 5% covered by living things. However, the composition of soil may vary from place to place. The properties of soil are determined by the composition of the soil, depending on different amounts of biotic and abiotic components.

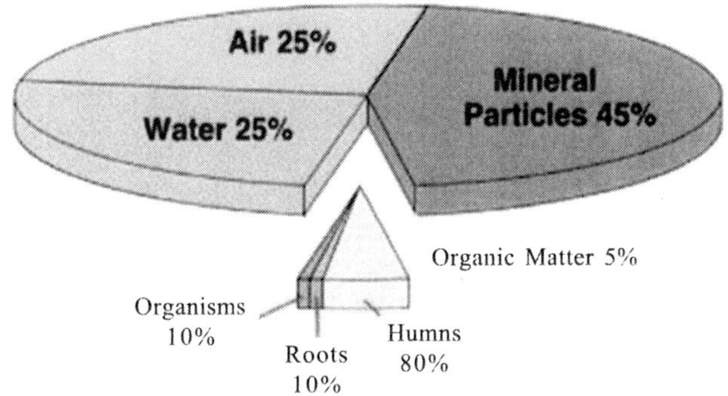

Fig. 4.3. General composition of soil

Soil has physical, chemical and biological properties. All these properties regulate the soil fertility or productivity.

Soil fertility is the ability of a soil to sustain plant growth by providing essential plant nutrients and favourable chemical, physical, and biological properties. Soil productivity is the capacity of the soil to produce crop with specific systems of management and it is expressed in terms of yields. All productive soils are fertile, but all fertile soils are not necessarily productive.

Physical properties: The physical properties of soil include soil texture, structure, density, porosity, consistency and colour. Soil physical environment is controlled by soil characters like texture, structure, aeration, water, mechanical resistance and depth of soil.

Soil texture refers to the size of soil particles dependent on proportion of sand, clay and silt.

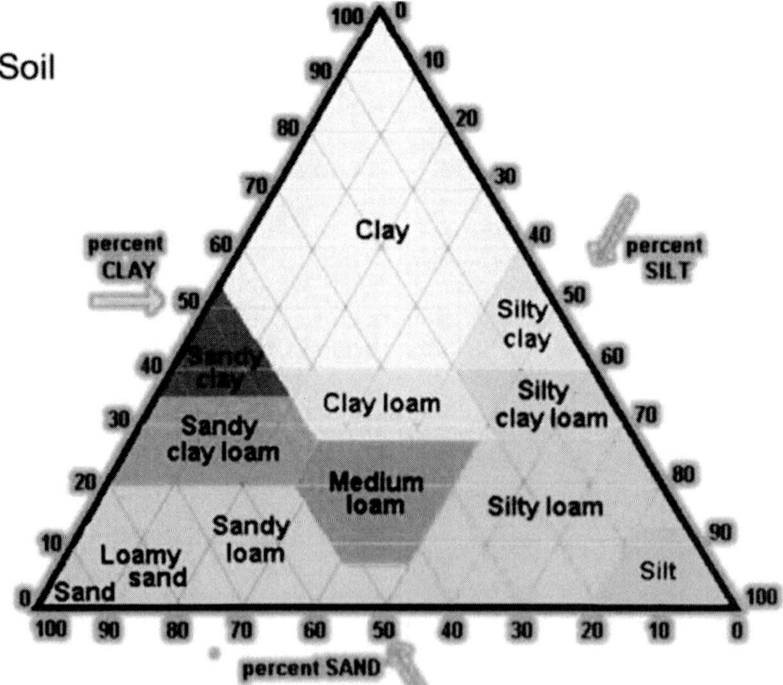

Fig. 4.4 Textural classification of soil

The textural components of soil, including sand, silt, and clay result in aggregates as a result of their clumping. The structure of the soil refers to the arrangement of such soil aggregates. Soil structure is influenced by physical processes that might be improved or destroyed by the choice of farming practices.

The average soil particle density ranges from 2.60 to 2.75 grams per cm^3 which usually remains unchanged for a given soil. It may change with the organic matter content of soil. Soil particle density is different from soil bulk density which is always less than soil particle density.

Soil porosity refers to the number of pores present within the soil that is determined by the movement of air and water within the soil. Soil porosity is important for moisture holding and drainage. Soil porosity is influenced by soil texture and structure.

Soil consistency is the ability of the soil to stick to it or other objects and to resist deformation and rupture. Depending upon the moisture level the consistency of dry soil ranges from loose to hard, whereas that of wet soil ranges from non-sticky to sticky.

Soil colour is determined primarily by the organic composition of the soil. Observation of soil colour is a qualitative means of measuring organic, iron oxide, and the clay contents of the soil.

Chemical properties: The soil chemical environment is dynamic and reactions that maintain dilute solution of nutrient element are indispensable for continual plant growth. The nutrient transformation and its availability in soils depend on pH, clay mineral, and cation and anion exchange capacity. Soil pH, cation exchange capacity (CEC) and salinity are usually considered as chemical properties of soil. The reaction of soil is expressed in terms of the soil pH, which determines the acidity and alkalinity of the soil. Soil pH is the measure hydrogen ion concentration in the aqueous solution of soil which ranges between 3.5 and 9.5. Usually, soils with high acidity contain higher amounts of aluminium and manganese, and soil with higher alkalinity has a higher concentration of sodium carbonate.

Cation exchange capacity is the maximum amount of total cations that a soil sample holds at a given pH. CEC is taken as an indicator of soil fertility.

Salts in the soil are transported from salt tables in water resources and accumulated on the surface due to evaporation. The salt accumulation affects the degradation of organic matter in soil and the vegetation on the soil.

Biological properties: The soil contains a vast array of life forms ranging from sub-microscopic (the viruses), to earthworms, to large burrowing animals such as gophers and ground squirrels through their burrowing and feeding activities which tend to improve aeration and drainage through structural modifications of the soil solum. In general, they affect soil chemical properties to a lesser extent though their actions indirectly enhance microbial activities due to creation of a more favourable soil environment. These microorganisms like bacteria (including actinomycetes), fungi, algae (including cyanobacteria) and protozoa play a vital role in degrading organic materials and regenerate supply of carbon dioxide for plants. Soil microbes exert much influence in controlling the quantities and forms of various chemical elements found in soil. Most notable are the cycles for carbon, nitrogen, sulphur and phosphorus, all of which are elements important in soil fertility, and as we know today, may be involved in global environmental phenomena. Mineralisation of organic matters by the soil microbes liberates carbon dioxide , ammonium, sulphate, phosphate and inorganic forms of other elements is the basis of nutrient cycle. Another important aspect of nutrient cycling is that under certain circumstances nitrogen and sulphur may be converted to gaseous forms (volatilized) and lost to the atmosphere. Nitrogen in the form of nitrate can be converted to gases such as nitrous oxide (N_2O) and di-nitrogen (N_2) through the process of denitrification. A consequence of denitrification is loss of nitrogen from the soil. On the other hand, this process is a useful way to remove excess nitrate from wastewater. In biological nitrogen fixation the bacteria and a few other microbes, notably the cyanobacteria (blue-green algae), atmospheric di-nitrogen (N_2) is captured and converted to plant-available forms.

Soil bacteria belonging to the genera *Rhizobium* and *Brady rhizobium* (and a few others) are capable of inducing the formation of nodules on roots of specific legumes (plants like peas, legumes etc.) fix large quantities of biological nitrogen by the process of symbiosis.

4.4 Plant nutrients

Plant contains small amount of 90 or more elements, but 17 elements are known to be essential. Three elements such as carbon, hydrogen and oxygen are supplied by air and water while remaining fourteen are supplied by soil.

Soils are the primary provider of nutrients and water for much of the plant life on earth in addition to providing anchorage to plants. It is generally accepted that there are 17 essential elements required for plant growth (Troeh & Thompson 1993). The lack of any one of these essential nutrients (Table 4.1), can limit the plant growth and development. The three non-mineral elements (C, H, O) are obtained from water and air while all other nutrients are supplied from mineral form. Three primary macronutrients (N, P, and K) are needed in the greatest quantities from the soil while three secondary macronutrients (Ca. Mg, S) are needed in smaller quantities. The micronutrients, or trace nutrients, are needed in very small amounts and excess quantity may be toxic to the plants.. Silicon (Si) and sodium (Na) are sometimes considered to be essential plant nutrients, but due to their ubiquitous presence in soils they are never in short supply (Epstein 1994, SubbaRao *et al.*, 2003).

Table 4.1. Essential elements for plants

Essential plant elements		Symbol	Primary form
Non-mineral elements (from air and water)			
	Carbon	C	$CO_{2\,(g)}$
	Hydrogen	H	H_2O (l), H^+
	Oxygen	O	H_2O (l), $O_{2(g)}$
Mineral elements			
Primary macro-nutrients	Nitrogen	N	NH_4^+, NO_2^-
	Phosphorous	P	HPO_4^{2-}, $H_2PO_4^-$
	Potassium	K	K+
Secondary macro-nutrients	Calcium	Ca	Ca^{2+}
	Magnesium	Mg	Mg^{2+}
	Sulphur	S	SO_4^{2-}
Micro-nutrients	Iron	Fe	Fe_3^+, Fe_2^+
	Manganese	Mn	Mn^{2+}
	Zinc	Zn	Zn^{2+}
	Copper	Cu	Cu^{2+}
	Boron	Bo	$B(OH)_3$
	Molybdenum	Mo	MoO_4^{2-}
	Chlorine	Cl	Cl^-
	Nickel	Ni	Ni^{2+}

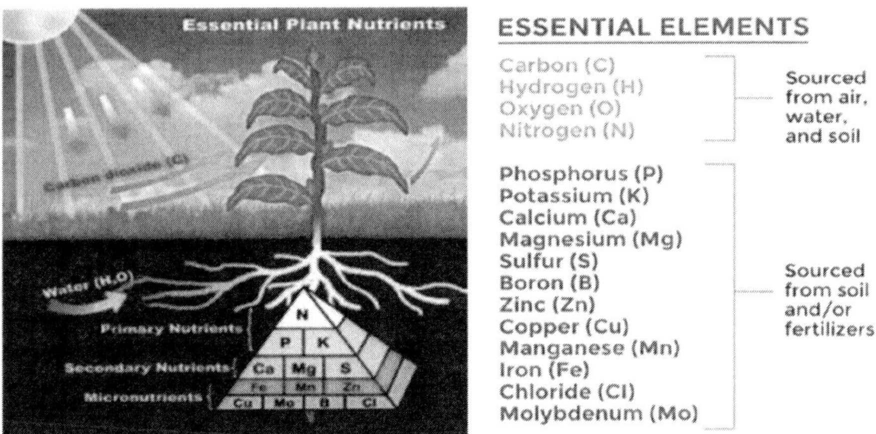

Fig. 4.5. Essential plant nutrients

Most of the essential nutrients are dissolved in soil water and electrically charged which are held by soil particles as they are also electrically charged. Soils also hold nutrients by retaining water. Clay and silt particles are the primary mineral components in soils that retain water. A major portion of water used by the plants are lost to the atmosphere by evapotranspiration stream while a small amount of it (less than 1%) is used for metabolic activities. Transpiration is a component of photosynthesis. These small particles slow the drainage of water and, like a sponge, physically hold water through capillary forces. Clay provides such strong force that plants can't pull all the water away from it, which makes silt particles the ultimate ingredient for plant-available water storage.

Soil, climate, crop variety and management factors exert considerable influence on plant composition. Because many biological and chemical reactions occur with fertilizers in soils, the quantity of nutrients absorbed by plants does not equal the quantity applied as a fertilizer. Proper fertilizer management can maximize the proportion of fertilizer nutrient absorbed by the plant. As plants absorb nutrients from the soil, complete their life cycle and die, the nutrients in the plant residue are returned to the soil. These plant nutrients are subject to the same biological and chemical reactions as fertilizer nutrients. Although this cycle varies somewhat among nutrients, understanding nutrient dynamics in the soil plant atmosphere system is essential to successful fertilizer management.

4.2.1 Major elements

Nitrogen (N)

Nitrogen is a key element in plant growth. Some plants such as legumes fix atmospheric nitrogen in their roots; otherwise fertiliser factories use nitrogen from the air to make ammonium sulphate, ammonium nitrate and urea. When applied to soil, nitrogen is converted to mineral form, nitrate, so that plants can take it up. Nitrate is easily leached out of soil resulting acidification.

Phosphorus (P)

Phosphorus helps transfer of energy from sunlight to plants, stimulates early root and plant growth, and hastens maturity. The most common phosphorus source is superphosphate, made from rock phosphate and sulphuric acid.

Potassium (K)

Potassium increases vigour and disease resistance of plants, helps form and movement of starches, sugars and oils in plants, and can improve fruit quality. Potassium is low or deficient on many of the sandier soils and heavy potassium removal can occur on soils used for intensive grazing and intensive horticultural crops (such as bananas and custard apples). Muriate of potash (potassium chloride) and sulphate of potash are the most common fertiliser sources of potassium.

4.2.2 Secondary nutrients

Calcium (Ca)

Calcium is essential for root health, growth of new roots and root hairs, and the development of apical portion of leaves. Lime, gypsum, dolomite and superphosphate (a mixture of calcium phosphate and calcium sulphate) all supply calcium. Lime is the cheapest and most suitable. Dolomite is useful for magnesium and calcium deficiencies, but if used over a long period will unbalance the calcium/magnesium ratio. Superphosphate is useful where calcium & phosphorus are needed.

Magnesium (Mg)

Magnesium is a key component of chlorophyll, the green colouring material of plants, and is vital for photosynthesis (the conversion of the sun's energy to food for the plant). Magnesium deficiency occurs in acid soils in high rainfall areas. Magnesium deficiency can be overcome with dolomite (a mixed magnesium-calcium carbonate), magnesite (magnesium oxide) or epsom salts (magnesium sulphate).

Sulphur (S)

Sulphur as a constituent of amino-acids is required for protein formation and is involved in energy-producing processes in plants. It encourages flavour and odour in plants and its products.

Sulphur deficiency is not a problem in soils high in organic matter, but it leaches easily. Superphosphate, gypsum, elemental sulphur and sulphate of ammonia are the main fertiliser sources.

4.2.3Trace elements/micronutrients

Iron: Iron is a constituent of many compounds of enzymes that regulate and promote growth and is readily available in acid soils.

Manganese: Manganese helps photosynthesis. It is freely available in acid soils, often in toxic amounts in very acid soils, but can be deficient in sandy soils. Toxicity is remedied with lime.

Copper: Copper is an essential constituent of enzymes in plants and can be deficient in organic soils. Overuse of another trace element, molybdenum, can cause copper deficiency in animals.

Zinc: Zinc helps in the production of a plant hormone responsible for stem elongation and leaf expansion. It is readily available in acid soils, but combines easily with iron in red soils. This is easily cured with the addition of zinc sulphate or crushed zinc minerals.

Boron: Boron helps with the formation of cell walls and translocation of sugars in rapidly growing tissue. Boron deficiency can reduce uptake of calcium in plants. It is chronically deficient soils used for horticulture but this is easily remedied with borax applied to the soil.

Molybdenum: Molybdenum helps bacteria and soil organisms of legumes to convert nitrogen in the air to soluble nitrogen compounds in the soil, so is particularly needed by legumes. The deficiency of molybdenum is found in acid soil, which can be corrected by application of sodium or ammonium molybdate.

Chlorine: Essential for photosynthesis and as an activator of enzymes involved in splitting water. It is associated with osmoregulation of plants growing on saline soils.

Nickel: It is associated with several enzymic systems. It contributes to nitrogen fixation and the metabolism of urea and is important for seed germination. Nickel favours growth of bacteria and fungi that are also required for plant growth.

The movement of essential nutrients in soil and key functions are presented in Table 4.2.

Table 4.2. Mobility and functions of essential plant nutrients

Essential nutrient	Mobility	Key functions
Nitrogen	Good	Proteins, protoplast, enzymes
Phosphorus	Good	ATP, ADP, Basal metabolism
Potassium	Good	Water and energy relations, cold hardiness
Sulphur	Fair/Good	Proteins, protoplast, enzymes
Calcium	Very poor	Cell structure, cell division, cell elongation
Magnesium	Good	Chlorophyll, enzymes
Boron	Very poor	Sugar translocation, cell development, growth regulation
Chlorine	Good	Photosynthesis
Copper	Poor	Enzyme activation
Iron	Poor	Chlorophyll synthesis, metabolism, enzyme activation
Manganese	Poor	Hill reaction- photosystem, enzyme activation
Molybdenum	Poor	Nitrogen fixation, nitrogen use
Zinc	Poor	Protein breakdown, enzyme activation
Nickel	Unknown	Iron metabolism

4.2.4 Non-essential elements

Nutrients which enhance the health but whose deficiency does not stop the life cycle of plants include: cobalt, strontium, vanadium, silicon and sodium. As their importance is evaluated they may be added to the list of essential plant nutrients, as is the case for silicon.

4.2.5 Sources of nutrients

Plant nutrients usually come from organic and inorganic sources. Organic nutrient sources are called bulky manures, organic manures or organic fertilisers.

Most organic nutrient sources, including waste materials, have widely varying composition and often only a low concentration of nutrients, which differ in their availability. Cereal straw has wide C:N ratio for which nitrogen release is very slow. The nitrogen-rich leguminous green manure and oilcakes decompose quickly and release nutrients swiftly. Crop residues and residues from animals are very important as they save nutrients due to recycling. Wastes obtained from plants and animals including industry are recycled to supply plant nutrients. A significant amount of N is made available through Biological Nitrogen Fixation (BNF) by a number of micro-organisms in soils either independently or in symbiosis with certain plants. The inocula of such micro-organisms are commonly referred to as bio fertilizers, which enrich soil by supplying biological nitrogen. However, organic sources supply limited quantity of nutrients as compared to inorganic sources. Inorganic fertilisers supply plant nutrients in larger quantity. In spite of less nutrient supplying quality, the organic sources have beneficial properties like physio-chemical and biological that help soil fertility. Rain water and irrigation water also supply some quantities of plant nutrients.

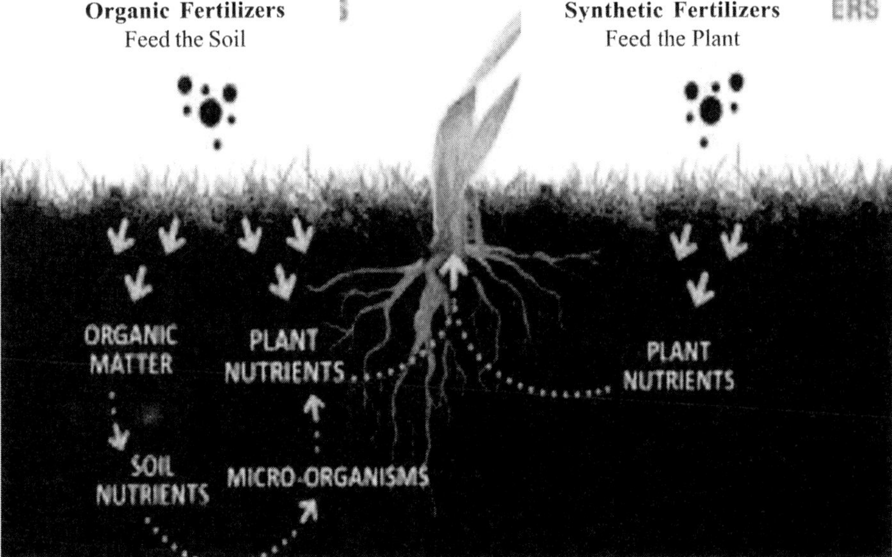

Fig. 4.6. Sources of nutrients

Nutrient content of some organic manures are given in Table 4.3.

Table 4.3. Nutrient content of some organic manures

Name of organic manure	Nutrient content in percentage		
	N	P_2O_5	K_2O
Cattle dung	0.35	0.12	0.17
Farm Yard Manure	0.75	0.20	0.50
Rural compost	0.75	0.20	0.50
Urban compost	1.75	1.00	1.50
Neem cake	5.22	1.08	1.48
Sugarcane trash	0.40	0.18	1.28
Pressmud	2.10	4.40	0.80

4.3 Soil health

Soil health needs to be judged from physical, chemical and biological properties of soil. A healthy soil can sustain plant and animal productivity and soil biodiversity, maintain or enhance water and air quality, and support human health and wildlife habitat. Soil is not just a medium to grow plants, but rather a living, dynamic and changing environment that is influenced by human activities. A healthy soil has physical, chemical, biological, microbiological and biochemical, and faunal indicators.

Apart from minerals, water, air and soil are living resource and home to more than 25% the planet's plant life and 95% of our food comes from the soil. Soil health influences quality and quantity of fruits, vegetables and food-grains. Soil helps fight climate change and global warming as well. In order to avoid soil pollution, we need to avoid single-use plastic and choose eco-friendly cultivation practices, proper water and nutrient management practices with minimum tillage. We should dispose of hazardous wastes like batteries responsibly and convert our food wastes and kitchen wastes into compost.

Management of soil health

Soil health management is the key to any agricultural production system. Soil degradation is a rapid process whereas soil formation is a very slow process. Depending on the soil forming process, it may take between 100 to 10,000 years for an inch of soil to form. Poorly managed soils can lead to unsustainable crop production and cause contamination of surface water and groundwater due to agrichemicals attached to the dust and soil.

There are no such universal indicators for soil health measurement. However, both quantitative and qualitative indicators can be worked out for different agro-ecosystems for measurement of soil health. Such indicators could be physical, chemical and biological.

There are many options and practices for management of soil health. These practices are (i) reduced tillage, (ii) Crop rotation, (iii) cover cropping, (iv) diversification of production system, (v) addition of organic matter and amendments, (vi) integration of crop-livestock production system, (vii) promotion of diverse plant species with different rooting system, and (viii) sustainable grazing practices.

Practices like soil conservation systems, including no-tillage and crop rotations can mitigate negative effects on soil health. A no-till system can restore soil health over time by improving soil infiltration, organic matter, water storage, soil structure, etc., which are indicators of soil health. Crop rotation and cover cropping increase bio-diversity and soil health.

4.4 Conservation of soil biodiversity

Soil biodiversity refers to the variety of organisms which live in the soil, including bacteria, fungi. It is the key ingredient that determines the fertility and productivity of land. Soil biodiversity determines carbon, nitrogen and water cycles, and the organic carbon present in soils is recognised as a major determinant of agricultural productivity. The amount of soil organic carbon is used as an indicator to track progress towards Sustainable Development Goals, which aims to achieve a land degradation-neutral world by maintaining healthy and productive land resources.

Soil biodiversity and soil health can be seen as one measure of environmental quality of soil from ecological point of view. In addition to protection of individual species, the focus has been shifted to conservation of habitats. To overcome any limitations to agricultural production, a series of potential "entry points" are needed at which management practices could be improved. These include both direct interventions such as: inoculation for disease and pest control and soil fertility improvement (such as rhizobia, actinomycetes, mycorrhizae, diazotrophs) and indirect interventions through cropping system design, organic matter management and genetic control of soil function.

4.5 Nutrient management of crops

There is a growing concern on contamination of environment leading to detrimental human and animal health due to application of chemical fertilisers. Very often people recommend application of organic nutrients to maintain the soil health and animal health including protection of environment. Scientists have found that there is emission of methane gas, carbon dioxide and nitrous oxide from rice fields, organic manures like cattle manure, burning of fossil fuels and human interventions, deforestation etc., that increase greenhouse gas (GHG) level of the atmosphere. Emission of GHG is a major cause of climate change. We have no option but to rely on increasing crop production vertically from the shrinkage land area to meet the food and nutrition requirement of burgeoning population. And increased production is not possible without application of essential plant nutrients.

All plants require certain nutrients for their survival, growth, development and reproduction. When such nutrients are found insufficient in the soil, additional nutrients are applied to soil for a good yield. These nutrients are added to the soil from commercial fertilizers or from organic sources such as manure, compost or bio-solids. Nature has a mechanism of nutrient cycles like nitrogen, phosphorous, potash, sulphur cycles that take care of soil nutrients. In spite of all these, plant growth is impaired in absence of certain elements which can be corrected by such specific elements only. Deficiency in any of the 17 essential nutrients has the potential to cause a decrease in crop quality or yield whereas excess of any nutrient may cause toxicity.

Nutrient management involves using crop nutrients as efficiently as possible to improve productivity while protecting the environment. The key principle behind nutrient management is balancing soil nutrient inputs with crop requirements and production target. However, the nutrients must be applied in right quantity at right time and in right manner. When excess nutrient is applied to soil or the applied nutrients are not effectively used by the plants, they may leach into ground water or surface water bodies. For example, too much nitrogen or phosphorus can impair water quality. The major focus of nutrient management should be to apply required nutrients for optimum yield and protect the environment.

We have to first determine what nutrients are in the soil (soil-testing) and what's available in a growing or harvested crop, and then determining what has to be added to meet the needs of crops. This nutrient management plan will lay out how nutrients are managed according to land base characteristics, crops being grown, type of nutrient, proximity to water and application methods. In inefficient nutrient management valuable nutrients could be lost, resulting in reduced crop yields or additional costs for commercial fertilizers. Site specific nutrient management has been displayed in Fig. 4.7.

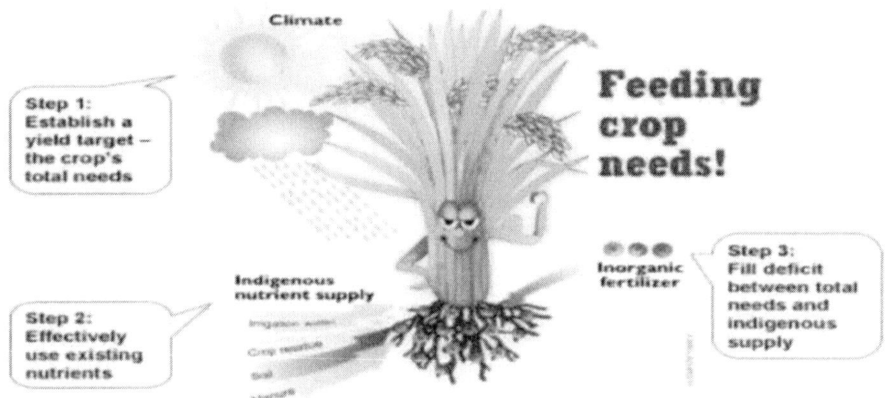

Fig. 4.7. Site specific nutrient management

Role of chemical fertiliser

The purpose of fertilizer use is to supplement the natural supply of soil nutrients, build up soil fertility in order to satisfy the demand of crops, and to compensate for the nutrient removal by crops from the soil. Since most people in developing nations are suffering from "hidden hunger" due to deficiencies of micro-nutrients such as iron, zinc and iodine, nutrient-enriched fertilisers are considered as a practical way to fight malnutrition.

In addition, micronutrient fertilization can extend the lifecycle of food, improve the post-harvest integrity of crops and thus reduce food waste: this is the case for calcium-based and boron-supplemented fertilizers, which help strengthen plant cells and make them more resistant.

Alternative to fertiliser

Organic manures can improve physical, chemical and biological properties of soil in addition to supplying plant nutrients to soil. Livestock manure is a valuable nutrient source, yet its nutrient content of manure varies widely between sources and farm management practices. Green manure is very efficient in crops like paddy. Crop residues (such as leaves, stems and roots) release the nutrients they contain when left on/in the soil. Crop residues vary greatly in nutrient content. Organic manure, farm yard manure or compost can be added to soil to meet the nutrient need of crops and act as soil conditioner. Biological N fixation (BNF) is the conversion of inert atmospheric nitrogen molecules (N_2) into forms of N that can be utilized by plants. Biological nitrogen fixation is done by leguminous crops at the site of root nodules by symbiotic relation by Rhizobium bacteria. Biological nitrogen fixation capacity of various leguminous crops range from 20 to 400 kg N/ha/year depending on plant species, length of the growing season and climatic conditions. Bio- fertilisers like Rhizobium, phosphorus solubilising bacteria (PSB), Azotobacter and Azospirillum are used to enhance biological fixation and mobilisation of nutrients.

Crop response

It is important to note that crops respond to different plant nutrients from all sources but they can take up nutrients only in their inorganic form. Therefore, the organic nutrients must be converted to inorganic form by mineralisation. Crop response to fertiliser depends upon the type of nutrients, availability of nutrient in soil and other conditions.

It is therefore, suggested that fertiliser application is to be made as per soil testing report and targeted crop yield.

Nutrition sensitive agriculture (NSA)

Malnutrition and undernourishment in children has been a regular feature in developing countries. Many children and women in reproductive stage suffer from anaemia. Further, according to the report of FAO, 34.7% of the children aged under five in India are stunted (too short for their age), while 20% suffer from wasting, meaning their weight is too low for their height. Malnourished children have a higher risk of death from common childhood illnesses such as diarrhoea, pneumonia, and malaria. The Global Hunger Index 2019 ranks India at 94 out of 102 countries on the basis of three leading indicators such as inadequate food supply, child mortality rate (under five) and child under nutrition (stunting, wasting). It is clear from the above data that India has not yet achieved food and nutrition security.

Food and nutrition security is achieved when people have physical, economic and social access to food of sufficient quantity in terms of variety, diversity, nutrient content and safety to meet their dietary needs and food preferences for a healthy and active life coupled with sanitary environment, adequate health and education care. In order to achieve food and nutrition security there is a need for transitioning toward nutrient-sensitive food system. Food systems are involved in the production, distribution, and consumption of food. A nutrition-sensitive approach not only considers policies related to macro-level availability of and access to nutrient-dense food, but it also focuses on household- and individual-level determinants of improved nutrition. Making the whole food system more nutrition sensitive requires a deliberate policy-oriented approach, which includes a combination of nutrition-specific as well as nutrition-sensitive interventions. More diverse diets are balanced in calorie, protein, and micronutrient intakes that lead to better anthropometric outcomes for all age groups and better overall cognitive outcomes.

As agriculture forms a major part of the food system, its role in enhancing nutrition always takes centre stage .Over the past four decades, the growth rate has played a key role in enhancing agricultural productivity as well as in raising incomes and kick-starting the structural transformation process.

Meeting the demand for staple grains has become the primary challenge .With the rise of income level of people .increased demand for diet diversification led to the rising demand for non-staple foods, such as vegetables, fruit, livestock, and dairy products. However, weak supply responsiveness for non-staples and their high relative prices, due to the persistence of growth rate-era policies that favour staple grains, resulted in limited access for the poor to a more nutritious diet which has caused malnutrition and under-nutrition. Therefore, policy focus on enhancing the diversity of the food system, particularly for micronutrient-rich horticultural and livestock products, is extremely crucial in these contexts.

Nutrition sensitive agriculture can be implemented in areas like (i) making food more available and accessible, (ii) making food more diverse and production more sustainable through diversification, and (iii) making food more nutritious by bio-fortification. Micronutrient-fortified staples, such as zinc-fortified rice and wheat, can play an important role in environments where markets are not well developed and where rural populations largely consume what they produce. Fortified non-staples, such as pulses, could be an effective source of micronutrients in areas with poor or better market infrastructure conditions.

The first pathway between agriculture and nutrition is through the income mechanism where gains in household income lead to increased access to food diversity. The second pathway explores all of the behavioural traits that impact nutrition such as the ability of the people to translate food intake into absorbed nutrients to achieve positive outcomes by safe drinking water and sanitation. Food-based safety nets like PDS are some of the most popular and widely used policy tools to dampen the effects of high and/or volatile food prices. They are an important part of the food distribution network that ensures timely access and availability of food grains at subsidized prices for the poor.

The success of a diversified food system in ensuring positive nutrition outcomes depends on intra-household equity in food access and an individual body's ability to convert food to nutrients. The most important among them are disease prevalence, water/sanitation facilities, and food safety. Water, sanitation, and hygiene (WASH) policies have a critical role in determining the health status of children. Traditionally, the distribution of food within the household has been lopsided in favour of men and older boys. This has meant that even if the household as a whole is food secure, there might be individuals (mostly women and girls) who might be food insecure. Therefore, enhanced emphasis on equitable distribution of food within the household, both in terms of quantity and quality, is extremely vital in determining individual-level nutrition.

4.6 Integrated soil fertility management (ISFM)

Vanlauwe et al. (2010) define ISFM as "a set of soil fertility management practices that necessarily include the use of fertilizer, organic inputs and improved germplasm, combined with the knowledge on how to adapt these practices to local conditions, aiming to maximize agronomic use efficiency of the applied nutrients and improved crop productivity". By definition, ISFM prescribes that interventions have to be aligned with prevalent biophysical and socio-economic conditions at farm and plot level (Vanlauwe et al., 2014). ISFM plays a critical role in both short-term nutrient availability and long-term maintenance of soil organic matter and sustainability of crop productivity in most smallholder farming systems in the tropics. The results of researches showed that the integrated application of organic and inorganic fertilizers improve productivity of crops as well as the fertility status of the soil. Combining fertilizers and organic inputs also enhances fertilizer uptake and retention by balancing immobilization and release processes (Chivenge et al., 2009).

Integrated crop nutrition

Integrated crop nutrition involves use of both organic and inorganic fertilisers. They can reduce need for both organic and inorganic fertilizers which have synergistic effect on each other. Organic manure improves soil physical, chemical and biological properties compared to inorganic fertilizers as indicated by reduction in bulk density, temperature and conservation of soil moisture. Organic wastes that were found to be effective in increasing soil nutrients contents, pH, nutrient uptake by crops and were successfully combined with inorganic fertilizers such as NPK.

4.7 Integrated nutrient management (INM)

Integrated Nutrient Management (INM) aims at integrating the use of all natural and man-made sources of plant nutrients for maintaining crop productivity without compromising the soil productivity of future generations. The INM philosophy combines economic and efficient traditional and improved technologies for symbiotic and synergetic crop-soil-environment bio-interactions. The basic concepts of IPNS is the maintenance or adjustment of soil fertility and supply of plant nutrients to an optimum level for sustaining desired crop productivity through optimization of benefits from all possible sources of plant nutrients in an integrated manner. INM aims to optimize the condition of the soil, with regard to its physical, chemical, biological and hydrological properties, for the purpose of enhancing farm productivity, whilst minimizing land degradation. Additionally, it is a method and a way of disposing organic wastes safely and also an effective method of recycling wastes into good quality compost.

In INM both organic and inorganic sources of nutrients are applied although balanced application of appropriate fertilizers is a major component of INM. However, fertilisers are required to be applied for optimal plant growth based on the requirement of the crop, quantity available from soil sources and agro-climatic considerations. Over-application of fertilizers, while inexpensive for some farmers in developed countries induces neither substantially greater crop nutrient uptake nor significantly higher yields (Smaling and Braun 1996). It is also not desirable for environmental point of view. Under application of fertiliser can retard crop growth and reduce crop yields. Balanced fertilization should also include secondary nutrients and micronutrients, in addition to macro-nutrients. Organic manures such as animal and green manures also aid soil conservation by improving soil structure and replenishing secondary nutrients and micronutrients (Kumwenda et al. 1996). Application of both organic and inorganic fertiliser not only conserves nutrients in the soil, but makes nutrient uptake more efficient. There are many reasons for loss of nitrogen in soil. Deep placement of fertilizers in soil provides a physical barrier that traps ammonia. The use of inhibitors or urea coatings that slow the conversion of urea to ammonium can reduce the nutrient loss that occurs through leaching, runoff, and volatilization. With innovations of these kinds, better timing, and more concentrated fertilizers, nutrient uptake efficiency can be expected to improve by as much as 30 per cent in the developed world and 20 per cent in developing countries by the year 2020 (Bumb and Baanante 1996).

The idea of INM depends on a number of factors, including harmony in nutrient properties, a balance between crop nutrient demands, what sort of nutrient, in general, is available in soil and in the farmer's hand, information and skills about the most suitable nutrient can be harmonized in combination, and which materials can be safely used that lead to increase nutrient-use efficiency. Additionally, it is a method and a way of disposing organic wastes safely and also an effective method of recycling wastes into good-quality compost (Selim et al, 2017, Roy et al, 2006).

INM is a simple system that can create a favourable soil condition, able to provide plants with sufficient, efficient, and sustainable nutrient source and also is a promising strategy that has made considerable contributions to lessen the negative environmental impact, boost both the quantity and quality traits of the global food supplies, and increase the land expansion with the plan of sustainable and economical agricultural development (Wu et al, 2014: Mueller et al, 2012).

Components of INM

The INM includes recycling of organic wastes, biological nitrogen fixation, green manuring and need based inorganics in an integrated manner. In addition soil amendments are used for correction of problematic soils. The components of INM system also include suitable crop variety, use of optimum cultural management, soil and water use for efficient and suitable crop production.

Organic manures: Organic manure is one of the major and commonly used organic nutrient components in INM. Organic manures include farm yard manure, compost, vermi-compost, poultry manure, sheep manure, night soil, oil cakes and excrete of other animals. The nutrient value of organic manures is not comparable to inorganic fertilizers. However, poultry manure/vermi-compost/ oil cakes are comparatively richer sources of nutrients and all of them play a vital role in maintaining soil fertility though their effects on physico-chemical properties of soil.

Compost is an amorphous brown to dark humified material produced as a result of microbial decomposition of organic wastes collected from urban and rural wastes. In addition to microbial decomposition, machines are also used to produce compost, commonly known as mechanical compost. Soil invertebrates (earth worms) are also used effectively for recycling of non-toxic/degradable organic wastes to the soil. Culturing of earth worms is referred as vermi-culture and the recycled granular produce is referred as vermi-compost. Composts are rich sources of essential plant nutrients. Besides, nutritional richness, composts are known to improve the physical, chemical and biological properties of the soil. Composts are also helpful in reducing the outlay on fertilizers.

Bio-fertilizers: Blue Green algae (BGA), Azolla, Rhizobium culture, Azotobacter, Azospirillum, Phosphorus Solublising Bacteria (PSB), Asperigillus niger, VAM etc. are some of the important bio-fertilisers applied in crop plants for nitrogen fixation or increasing availability of phosphorus. Biofertilizers are the products containing living cell organisms capable of fixing atmospheric nitrogen, solubilizing and mobilising phosphorus in soil, produce growth, promoting and antifungal substances. Biofertilizers are either symbiotic (Rhizobium) or non-symbiotic (Azotobactor and Azospirrilium). These biofertilizers are used as seed, seedling and soil inoculant

Chemical fertilizer : Urea, Urea super granules, Single Super phosphate, Muriate of potash, micronutrient fertilizers are mostly used in the crop fields.

Green manure and Green leaf manure: Crops grown for restoring or increasing the organic matter content in the soil are referred as green manure crops and this cropping system is called as green manuring. Crops here are grown either in situ or brought from outside and upon decomposition, besides releasing nutrients; they add organic matter, produce enzymes, vitamins and antibiotics. *Sesbania rostrata*, *S. aculeata*, *Crotalaria juncea*, Azolla, etc. are commonly used green manures. *Sesbania rostrata* (stem nodule) can fix 100-250 kg N ha^{-1} in 40-45 days in rice crop.

Crop rotation: Crop rotation is beneficial in sustaining both yield and quality of crops, as it has a potential to overcome all those factors which are responsible for decline in yield like loss of soil fertility, unbalanced nutrient uptake & presence of pest and weeds, legumes in rotation have been found to increase the yield by 25-30%, besides fixing atmospheric nitrogen to the extent of 30-40 kg per ha in tropical and sub-tropical regions.

Crop residues: Crop residues are non-economical plant parts that are usually left in the field after harvest, left in packing sheds or processing units and serve as a potential source of nutrients, besides promoting and improving soil and water conservation, soil fertility and crop productivity.

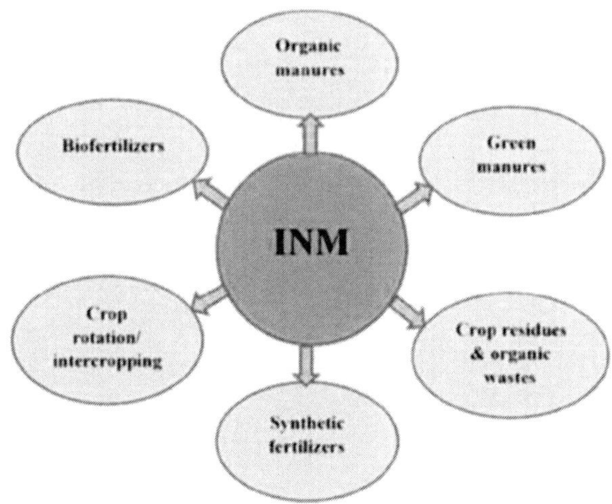

Fig. 4.8. Components of INM

References

Bannick, C, Mitt. Dtsch. Bodenk. Ges. 73 (1994) 23 –26

Blumes, HP (1990) : Handbuch des Bodenschutzes, ecomed, 2nd ed., Landsberg 1990

Bosch, C (1988) in D. Rosenkranz, G. Bachmann, G. Einsele, H.- M. Harreß (eds.): Bodenschutz, Erich Schmidt Verlag, Berlin 1988, p. 1480

Bumb, B., and C. Baanante. 1996. The role of fertilizer in sustaining food security and protecting the environment to 2020. 2020, Vision Discussion Paper 17. Washington, DC: IFPRI.

Chivenge P, Vanlauwe B, Gentile R, Wangechi H, Mugendi D, van Kessel C, Six J, 2009. Organic and mineral input management to enhance crop productivity in Central Kenya. Agronomy Journal 101, 1266-1275.

Haygartham PM, Karl Ritz (2009): The future of soils and land use in the UK: Soil systems for the provision of land-based ecosystem services

Klapp, E. (1958): Lehrbuch des Acker- und Pflanzenbaus, Paul Parey, Hamburg 1958.

Kumwenda, J. D. T., S. R. Waddington, S. S. Snapp, R. B. Jones, and M. J. Blackie. 1996. Soil fertility management research for the maize cropping systems of smallholders in southern Africa: A review. Natural Resources Group Paper 96-02. Mexico City: International Maize and Wheat Improvement Center (CIMMYT)

M. M. Selim and A.-J. A. Al-Owied, "Genotypic responses of pearl millet to integrated nutrient management," *Bioscience Research*, vol. 14, no. 2, pp. 156–169, 2017.

N. D. Mueller, J. S. Gerber, M. Johnston, D. K. Ray, N. Ramankutty, and J. A. Foley, "Closing yield gaps through nutrient and water management," *Nature*, vol. 490, no. 7419, pp. 254–257, 2012

R. N. Roy, A. Finck, G. J. Blair, and H. L. S. Tandon, *Plant Nutrition for Food Security*, FAO, Rome, Italy, 2006.

Smaling, E. M. A., and A. R. Braun. 1996. Soil fertility research in Sub-Saharan Africa: New dimensions, new challenges. Communications in Soil Science and Plant Analysis 27 (Nos. 3 and 4).

Vanlauwe B, Chianu J, Giller KE, Merck R, Mokwenye U, Pypers P, Shepherd K, Smaling E, Woomer PL, Sanginga N, 2010. Integrated Soil Fertility Management: Operational definition and consequences for implementation and dissemination. Outlook on Agriculture 39, 17-24. Vanlauwe B, Descheemaker K, Giller KE, Huising J, Merck

Vanlauwe B, Descheemaker K, Giller KE, Huising J, Merckx R, Nziguheba G, Wendt J, Zingore S, 2014. Integrated Soil Fertility Management in sub-Saharan Africa: Unravelling local adaptation. Soil 1, 1239-1286.

W. Wu, C. Li, B. Ma, F. Shah, Y. Liu, and Y. Liao, "Genetic progress in wheat yield and associated traits in China since 1945 and future prospects," *Euphytica*, vol. 196, no. 2, pp. 155–168, 2014.

5

Water Management and Dryland Technology

5.1 Water potential in India

Soil, climate and water decide the nature of crop to be grown in a particular region. While soil and climate can be manipulated to certain extent, there is no substitute of water. Water is governed by the hydrological cycle. The major source of water is rainfall which has been reduced over the years due to climate change. The per capita water availability in India was 4944 m^3 in 1955 which has been reduced to 1750 m^3 now and is expected to come down to 1500 m^3 by 2025.India gets about 400 m ha m of water annually of which 70 m ha m is lost by immediate evaporation and 115 m ha m runs off. Of the 215 m ha m of water that infiltrates into soil, 50 m ha m goes to ground water and 165 m ha m is stored at the root zone for crops. The ultimate irrigation potential of the country is estimated as 140 m ha m. It is estimated that 58.5 m ha m will be the potential from Major and Medium projects while Minor irrigation projects can contribute 81.4 m ha m. India is blessed with many rivers out of which 12 are major rivers with a catchment area of 252.8 m ha.

5.2 Soil-plant-water relationship

Soil is a dynamic body and mixture of minerals, organic matters, nutrients, living organisms, gases and water that support life system. It provides anchorage to plants and meets all needs of the plants that produce food by photosynthesis at the site of chlorophylls with the help of solar energy, carbon dioxide, enzymes and water.

5.2.1 Soil composition

On volumetric basis most soils are composed of about 50 per cent solids (45% minerals and 5% organic matter) and 50 per cent pore space occupied by air and water. The mineral matter consists of small particles of sand, silt, or clay. Organic matter can account for up to about 5 per cent of the overall soil makeup by volume, but many agricultural soils have less than 1 per cent organic matter. The pore spaces that occur between the mineral particles, are important because they store air and water in the soil. Air and water in the pore space are complementing each other. The amount of air in pore space decreases with the increase in amount of water and vice versa. Fig. 5.1 indicates the general composition of soil.

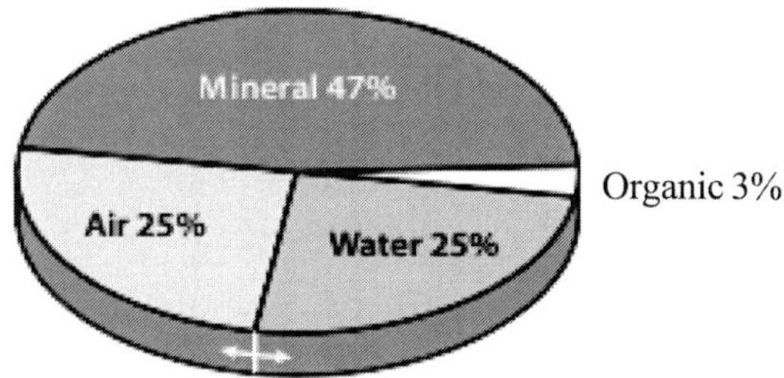

Fig. 5.1. A typical soil composition

5.2.2 Soil physical properties

Soil Texture: Soil texture refers to the sizes of the particles that make up the soil. The traditional method of determining soil particle size consists of separating the particles into three convenient size ranges. Thee three soil separates are sand, silt and clay. The soil separates are classified as per their size determined by USDA and ISSS (Table 5.1).

Table 5.1. Soil separates as per USDA and ISSS

Fraction	Particle diameter in mm	
	USDA	ISSS
Gravel	>2	>2
Very coarse sand	1-2	-
Coarse sand	0.5-1	0.2-2.00
Medium sand	0.25-0.5	-
Fine sand	0.1-0.25	0.02-2.00
Very fine sand	0.05-0.1	-
Silt	0.002-0.05	0.002-0.02
Clay	< 0.002	< 0.002

Clay is an important soil fraction because it has the most influence on soil behaviour such as water-holding capacity. Clay and silt particles cannot be seen with the naked eye. Texture is determined as per predominant size fraction. When the three fractions occur in equal proportion the soil is called "loam" .The soil textural triangle (Fig.5.2) shows the different textural classes and the percentage by weight of each soil fraction. For example, a soil containing 30 per cent sand, 30 per cent clay, and 40 per cent silt by weight is classified as a clay loam.

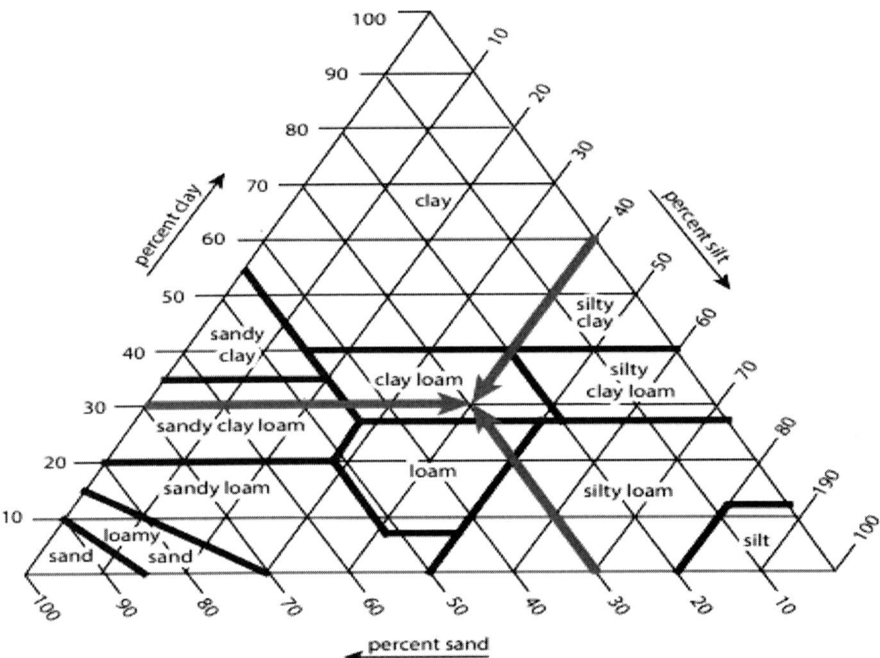

Fig. 5.2. Soil textural classification

'Soil colloids' refer to the finest clay and soil organic matter particles in a soil. Colloids are an important soil fraction as the sites for most physical and chemical activity in the soil. One such property is their large surface area. Smaller particles have more surface area for a given volume or mass of particles than larger particles. Thus, there is increased contact with other colloids and with the soil solution which results in the formation of strong friction and cohesive bonds between colloid particles and soil water. This is the reason why a clay soil holds together better than a sandy soil when wet.

Soil structure: The arrangement of individual soil particles with respect to each other in to a pattern is termed as soil structure. The binding of soil particles into larger clusters is called aggregates or 'peds. The major forms of soil structure are platy, prismatic, columnar, blocky, and granular. These soil structure descriptions indicate how the particles arrange themselves into aggregates. Structure less soils can be either single-grained (individual unattached particles, such as a sand dune) or massive (individual particles adhered together without regular cleavage, such as clay pans or hardpans). Aggregates vary according to their resistance to the impact of rain drops or under submergence. The stability of aggregates depends upon the clay content, the extent of flocculation, the organic-inorganic linkages, the microbial glue of the aggregates and the presence of mineral cementing material, such as iron and aluminium oxides. Soil structure has a strong effect on soil properties like erodibility, porosity, hydraulic conductivity, infiltration and water holding capacity.

Soil bulk density and porosity: The oven dry weight of a unit volume of soil inclusive of pore spaces is called bulk density. The bulk density of a soil is always smaller than its particle density. The bulk density of sandy soil is about 1.6 g / cm3, whereas that of organic matter is about 0.5. Bulk density normally decreases, as mineral soils become finer in texture. The weight per unit volume of the solid portion of soil is called particle density. Generally particle density of normal soils is 2.65 grams per cubic centimetre. With increase in organic matter of the soil the particle density decreases. Particle density is also termed as true density. The total volume includes both the solids and the pore spaces. Soil bulk density is important because it is an indicator of the soil's porosity.

Soil texture and structure influence porosity by determining the size, number and interconnection of pores. Coarse textured soils have many large (macro) pores because of the loose arrangement of larger particles with one another whereas fine-textured soils are more tightly arranged and have more small (micro) pores. Because fine-textured soils have both macro- and micro pores, they have a greater total porosity than coarse-textured soils. Long-term cultivation tends to lower total porosity because of a decrease in SOM and large peds (Brady and Weil, 2002).

5.2.3 Water relations of soil

Soil is a reservoir of water and plant nutrients that are required for the plants to grow. Water holding capacity of soil varies from soil to soil as per texture and organic matter content. If a sandy soil and clay soil having the same water content are placed in intimate contact with one another, water will move from the sandy soil to the finer textured soil. Here lies the importance of soil water potential to describe such a phenomenon.

Soil water: Soil water is usually described in terms of gravimetric per cent, per cent on a volume basis, or equivalent water depth per unit of soil depth. These are required for irrigation purpose when one has to decide how much irrigation water is required to bring the soil back to ideal water content. But it is also important to understand the behaviour of different soils that vary according to water content and why plants respond differently even though different soils have similar water contents. Soil water potential is the property used to describe such a phenomenon.

How soil holds water?: Soil holds water either as a thin film on individual soil particles or as water stored in the pores of the soil. Water held as a thin film by adhering to the outside layers of soil particles is called adsorption and soil water stored in pores is by capillary forces. Capillary rise of water can be explained the rise of water in a tube which depends on the diameter of the tube (Fig. 5.3).

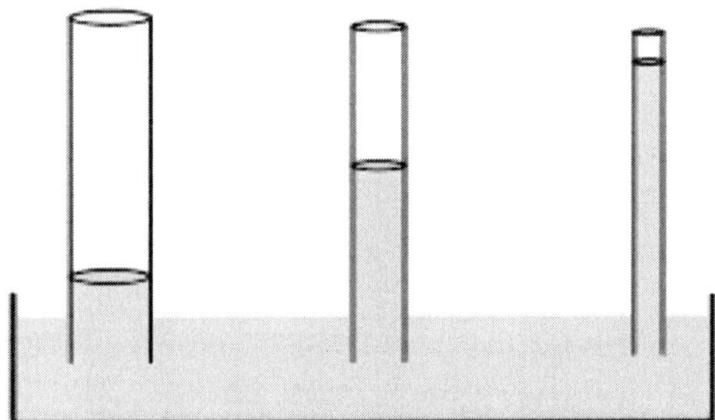

Fig. 5.3. Capillary rise of water in different tubes

Soil water tension: Soil water tension or potential is a measure of tenacity with which water is retained in the soil. The water potential is expressed in atmosphere or bar with respect to free water. Water is stored in pores by the capillary storage or by adsorption at a certain tension. As the soil dries, these tensions become larger. Plants can extract water easily when water is held in soil at a lower tension. At saturation, the soil water tension is approximately 0.001 bar. Field capacity is approximately one-third atmosphere pressure or approximately 0.3 bar. This stage is reached after saturation when the down word movement is virtually ceased and plants easily extract water from the soil. The wilting point is another stage which occurs when the potential of the plant root is balanced by the soil water potential and water is retained at – 15 bar approximately. Plant is unable to extract water from the soil at such tension and shows the symptoms of wilting. Prolonged exposure at this stage will cause death of plant. A soil water characteristic curve (Fig. 5.4) illustrates the tension relationship for different types of soils due to different soil textures and structures. Water between the field capacity and the wilting point is water relatively available to the plant. Best plant growth and yield for most field crops occur when the soil water content remains in the upper half of the plant available soil water range. This is called readily available water.

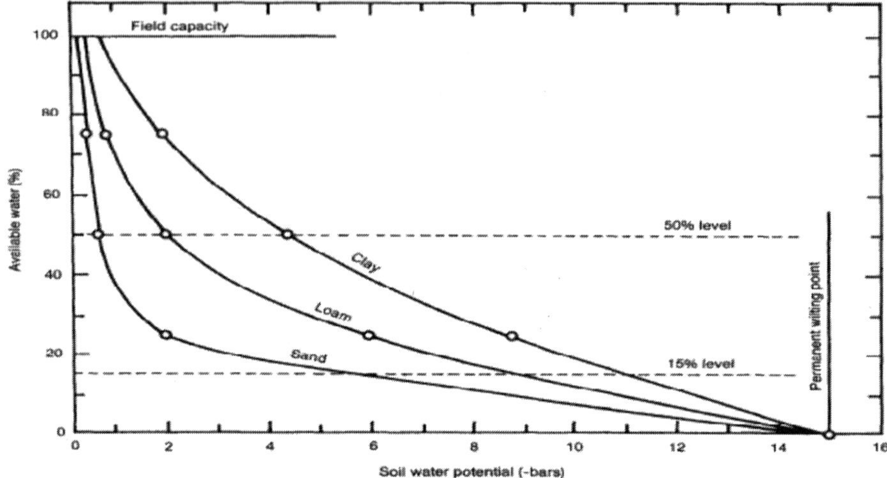

Fig. 5.4. Water relation curve for three soils

Soil water considered to be available for plant growth lies at a potential energy level between FC and PWP. But these determined values represent only the matric potential of the soil water system. The presence of salts may contribute a substantial osmotic component to the total soil water potential.

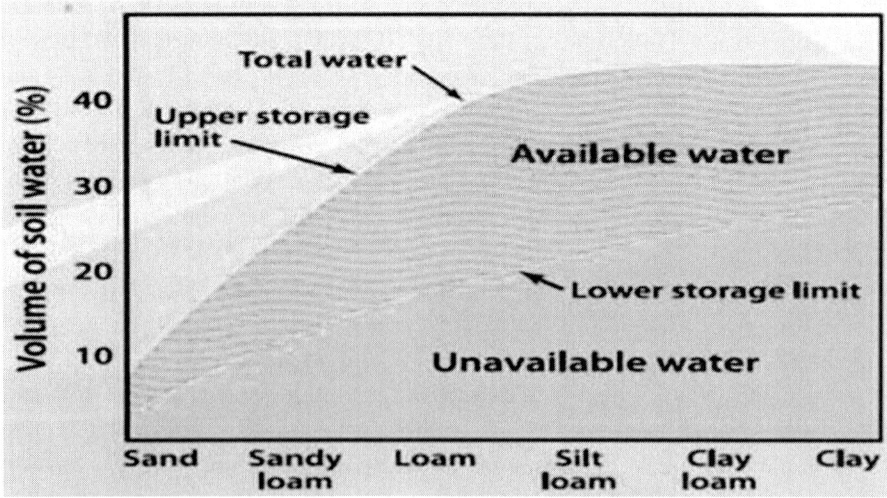

Fig. 5.5. Available soil water

Table 5.2. Typical FC and wilting point values (m^3 m^{-3}) for different soils

Soil textures	Field capacity	Willing point	Available water
Coarse sand	0.06	0.02	0.04
Fine sand	0.10	0.04	0.06
Loamy sand	0.14	0.06	0.08
Sandy loam	0.20	0.08	0.12
Light sandy clay loam	0.23	0.10	0.13
Loam	0.27	0.12	0.15
Sandy clay loam	0.28	0.13	0.15
Clay loam	0.32	0.14	0.18
Clay	0.40	0.25	0.15
Self-mulching day	0.45	0.25	0.20

(*Source:* Attila Yazar, 2013)

Total water potential consists of several components and it is the sum of matric potential, solute potential, gravitational potential and pressure potential. Gravitational potential is independent of soil properties and depends only on the vertical distance between the reference and the point in question. Matric potential is a dynamic soil property and will be at a theoretical zero level for a saturated soil. The matric potential of a soil system results from capillary and adsorptive forces due to the soil matrix. This potential was formerly called capillary potential or capillary water. Pressure potential applies mostly to saturated soils.

Where water quantity is expressed as a weight, pressure potential is the vertical distance between the water surface and a specified point. In the field, this component is zero above and at the level of water in the piezometer. Below the water level it is always positive.

Solute or osmotic potential arises because of soluble materials (generally salts) in the soil solution and the presence of a semipermeable membrane. Two recognized membranes in soil-water systems are the cell wall of plant roots and air-water interfaces.

Use of water by plants

A plant's root system must provide a negative tension (pressure) to extract water from the ground which needs to be equivalent to the tension that holds the water in the soil. Water movement takes place from higher water potential to lower water potential. When the plant can no longer extract water from the soil and will be under severe stress to the point of death. There are several factors responsible for water use by plants. These factors include daily plant water need as influenced by climatic conditions and stage of growth, plant root depth, and soil and water quality. Usually plants require water differently at different stages of growth. The water requirement of plant is less in early growth stages and it is more in reproductive stages.

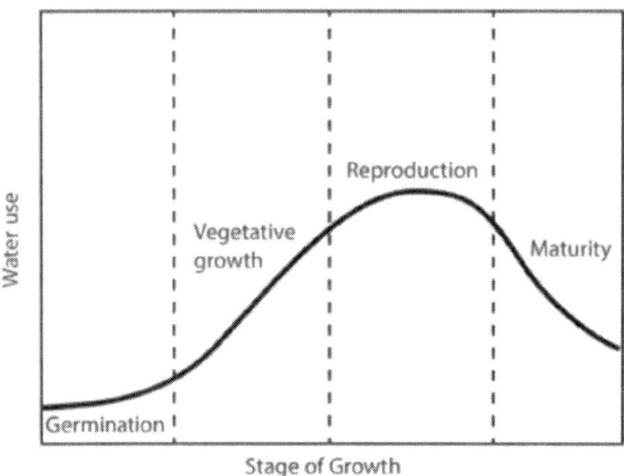

Fig. 5.6. Typical plant water need at different growth stages

Water extraction by plants also depends by their rooting depth. Normally water extraction is more from top quarter of rooting zone than the bottom quarter. A young plant with only shallow roots cannot have access to soil water from dipper zones. Plants typically extract about 40 per cent of their water needs from the top quarter of their root zone followed by 30 per cent from the next quarter, 20 per cent from the third quarter, and only 10 per cent from the deepest quarter (Fig. 5.7).

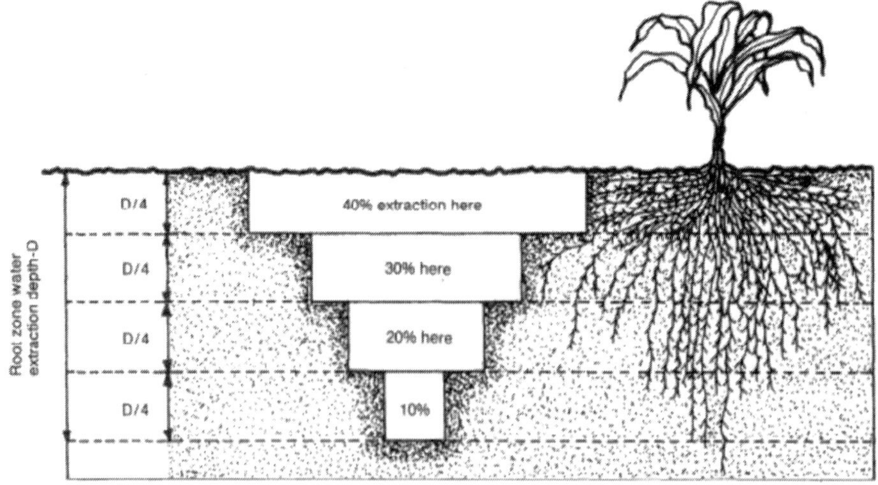

Fig. 5.7. Soil moisture extraction pattern from different rooting depth

Soil and water quality also influence plant water use. For good plant growth, a soil must have adequate room for water and air movement, and for root growth.

A soil's structure can be altered by certain soil management practices. For example, excessive tillage can break apart aggregated soil and excessive traffic can cause compaction. Both of these practices reduce the amount of pore space in the soil, reducing the availability of water and air, and reducing the room for root development. Irrigation water with a high content of soluble salt is not as available to the plant as increasing salt content of the water reduces the potential to move water from the soil into the roots. In such cases some additional water would also be needed to leach out the salt below the crop root zone to prevent salt build-up in the soil.

5.3 Irrigation scheduling

Penman (1948,1956) introduced the concept of potential evapotranspiration which was defined as 'the amount of water transpired in unit time by short green crop of uniform height completely covering the soil and never short of water'. Actual transpiration by the crop is a certain fraction of potential evaporation.

Evapo-transpiration or consumptive use is the quantity of water transpired by plants during their growth period or retained in the plant tissue plus evaporation from soil surface and vegetation. The consumptive use is the amount of water used for evapo-transpiration and water used for metabolic processes which is less than 1 per cent. The total amount of water used in evapo-transpiration by a crop during the entire season is called seasonal consumption of water. Evapo-transpiration is measured from evaporation data or climatological data. The water requirement of crops includes losses due to evapo-transpiration or consumptive use plus losses during application of irrigation water and water required for specific purposes like land preparation, transplanting , leaching etc.

Irrigation requirement of crops depends on climate, season, crop duration, variety and soil conditions. Scheduling irrigation is an attempt to irrigate at the appropriate amount for a specific stage in the plants development and growth. Irrigation schedule needs to indicate the quantity and time of applying irrigation water. The main objectives of efficient irrigation management are : (i) high yield of good quality, (ii) high water use efficiency, (iii) least damage to soil productivity, and (iv) low irrigation cost.

Several approaches for scheduling of water have been adopted by the scientists and farmers.

i) **Soil moisture depletion approach:** Usually irrigation should be given when about 50% of available soil moisture is depleted from the root zone. The available water is the soil moisture that lies between field capacity and wilting point. The relative availability of soil moisture is not the actual difference between field capacity and wilting point as the crop might suffer before the soil moisture reaches wilting point. Therefore it is necessary to locate the optimum point within the available range of soil moisture, when irrigation must be scheduled to maintain crop yield at high level.

ii) **Plant indices:** Deficit of water in the plant can be seen from symptoms like dropping, curling or rolling of leaves and change in foliage colour and this can be taken as an indicator for irrigation scheduling. However, these symptoms indicate the need for water without indicating the quantity to be applied. Growth indicators such as cell elongation rates, plant water content and leaf water potential, plant temperature leaf diffusion resistance etc. are also used for deciding when to irrigate. Some indicator plants like sunflower h is used for estimation of PWP of soil.

iii) **Climatological approach:** Evaopo-transpiration is approximately used as consumptive use, which is influenced by climatic factors. The amount of water lost by evapotranspiration is estimated from climatological data and when ET reaches a particular level, irrigation is scheduled. There are methods like modified Penman and pan evaporation from free water surface are used for calculation of PET. The ET_p is less than ET_0 for which crop coefficient is calculated through field research. Prihar et al (1974) suggested a more practical approach of IW/E_{pan} (The ratio between fixed amount of irrigation water $_{(IW)}$ and cumulative pan evaporation minus rain) as a basis for scheduling irrigation. The amount of irrigation given is either equal to ET or fraction of ET. In IW/CPE approach, a known amount of irrigation water (say 5 cm) is applied when cumulative pan evaporation (CPE) reaches a predetermined level. The amount of water given at each irrigation ranges from 4 to 6 cm as per soil texture. Scheduling irrigation at an IW/CPE ratio determined by field experiments for different crops and stages of growth as available (0.50, 0.60, 0.75, 0.80, 1.00) for different locations is usually given.

Table 5.3. Optimum IW/CPE ratios for scheduling irrigation in important crops

Crop	Optimum IW/CPE ratio
Groundnut	0.75 to 1.0 IW/CPE ratio depending on crop developmental stages in Andhra Pradesh, Maharashtra & West Bengal
Sunflower	0.5 to 1.0 IW/CPE ratio depending on crop developmental stages at Hyderabad & Kanpur
Wheat	1.0 IW/CPE ratio at Ludhiana, Kanpur and Bikramganj
Gram	0.4 IW/CPE ratio at Ludhiana
Mustard	0.4 IW/CPE ratio at Hissar
Maize	0.75 to 1.0 IW/CPE ratio depending on crop developmental stages at Delhi & Hyderabad
Sugarcane	0.5 to 1.0 IW/CPE ratio depending on crop developmental stages at Lucknow

(*Source:* http://www.angrau.ac.in/media/7380/agro201.pdf)

iv) **Critical growth approach:** There are some growth stages of a crop at which moisture stress leads to irrevocable yield loss. These stages are known as critical growth stages or moisture sensitive periods. Under limited water supply conditions, irrigation is scheduled at moisture sensitive stages and irrigation is skipped at non-sensitive stages. The critical crop growth stages of some important crops are indicated in Table 5.4.

Table 5.4. Moisture sensitive periods of some important crops

S.No.	Crop	Moisture sensitive stage
1.	Rice	Panicle initiation, flowering
2.	Wheat	CRI, jointing, milking
3.	Maize	Silking, tasselling
4.	Groundnut	Flowering, pegging, podding
5.	Green gram	Flowering, pod formation
5.	Sugarcane	Formative stage
6.	Cotton	Flowering, boll formation
7.	Chilli	Flowering
8.	Potato	Tuber initiation, tuber maturity
9.	Onion	Bulb formation, maturity
10.	Tomato	From commencement of fruit set

Real time irrigation (RTI) is broadly defined as application of irrigation water based on the actual measured value of soil moisture at the plant root level. Water is applied when the soil moisture is less than the optimum value. The application and duration of irrigation is hereby based on the true value of moisture at the time and climatic conditions. This requires the processing of climatological data day-by-day and will only be used under technological well advanced conditions and skilled farmer, which can monitor the soil moisture on a daily base and which can effect irrigation at irregular intervals (Abdulmumin et al., 1990).

No single method of irrigation scheduling can provide an accurate assessment of the needs and balances of the soil-plant-water system. It requires a combination of two or more methods. Field information of evaporative demand and soil and plant water status should be used in conjunction with a water budget approach to determine if scheduled applications are sufficient; eliminating unintentional leaching of plant nutrients (Cassel, 1984, Perrier & Salkini, 1987, Martin et al., 1990; Jones, 2004; Pereira et al., 2006.

5.4 On-farm water management

Efficient water management is very crucial as huge amount of water is lost during distribution. It is roughly estimated that hardly 20-40% of the water released from the major reservoir is available to crops and the rest is lost in one form or the other (Boss, 1977).Therefore on-farm water management is required for efficient irrigation.

On-farm water management is a system approach towards controlling water on a farm in a manner that provides for the beneficial management of water for satisfying the irrigation and drainage needs of a crop. Sufficient water must be present in the root zone for germination, evapotranspiration, nutrient absorption by roots, root growth, and soil microbiological and chemical processes that aid in the decomposition of organic matter and the mineralization of nutrients.

During crop evapotranspiration (ET) process, meteorological factors control the strength of the "sink", soil factors control the source of water and plant factors control the transmission of water from the source to the sink. Therefore the soil-plant-atmosphere continuum must be considered for irrigation. Accurate irrigation scheduling, based on sound scientific principles, is becoming more important now. Thus, the purpose of optimum irrigation scheduling is to ensure an adequate supply of soil moisture to minimize plant water stress during critical growth stages (Wanjura et al., 1990). To improve water use efficiency a coordinated approach at different levels of the irrigation system is required: at the water source, at the conveyance system and at farm and field level (Yazar et al., 1995). On-farm water management covers conveyance, irrigation, field application and drainage of irrigation water.

i) Conveyance

Conveyance structures transport water to and from various locations on the farm, and include laterals, manifolds, sub-mains, and mains. The primary purposes of these structures are to transport water from the source (either surface or ground water) to the irrigation system or from the field to the sink where runoff is deposited.

Earthen water courses, open lined channels and underground pipes are used to reduce seepage and conveyance loss.

ii) Methods of irrigation

There are many methods like surface and sub-surface irrigation. Surface methods include Border Strip method, Flooding, Check-Basin method, Furrow method, Sprinkler method and Drip method. Sub-surface irrigation methods include sub-irrigation. The water use efficiency is the highest in drip irrigation followed by sprinklers under surface irrigation.

Border irrigation: Border irrigation uses land formed into strips, bounded by ridges or borders The borders are divided by levees running down the slope at uniform spacing. In this method, water is applied at the upper end of the border strip, and advances down the strip. Irrigation takes place by allowing the flow to advance and infiltrate along the border.

Basin method: Basin irrigation is the simplest of the surface irrigation method where water is applied to levelled surface units (basins) which have perimeter dikes to prevent runoff and to allow infiltration. This method is the most commonly practiced worldwide, both for rice and other field crops, including orchard tree crops. In general, basins are small and uneven and water application is manually controlled. Basins may be of different types: rectangular, ring, and contour.

Furrow irrigation: A furrow is a small and shallow channel constructed down or across the slope of the field to be irrigated parallel to row direction. Furrow irrigation is defined as a partial surface flooding method of irrigation where water is applied in furrows or rows of sufficient capacity to obtain the designed irrigation system. It is suitable for row crops at certain spacing.

Sprinkler irrigation: In sprinkler irrigation, water is delivered through a pressurized pipe network to sprinklers, nozzles, or jets which spray the water into the air, to fall to the soil as an artificial. Under this micro-irrigation system water is distributed through a network of pipe lines to and in the field and applied through selected sprinkler heads or water applicators. The water use efficiency is about 70%.

Drip irrigation: In this system water is applied at constant steady flow to soil at low pressure directly to the root zone of plants by means of applicators (orifices, emitters, porous tubing, perforated pipe, etc.) which operates under low pressure with the applicators being placed either on or below the surface of the ground . Water loss is minimized through these measures, as there is very little splash owing to the low pressure and short distance to the ground. Drip systems tend to be very efficient up to 90% and can be totally automated.

iii) Drainage

Drainage is the removal of excess water from the root zone area which improves the soil as a medium for plant growth. The sources of excess of water may be precipitation, irrigation water, snowmelt, overland flow, artesian flow from deep aquifers, flood water from channels, or water applied for special purposes as leaching salts from the soil. The function of a drainage system is to remove excess water from a field root zone to avoid permanent root damage resulting from lack of oxygen in the soil. Both the surface drainage and underground drainage systems are used. Drainage systems need to be compatible with crops grown, field layouts, and cultural practices such as crop rotation and cultivation.

Drainage improves the soil condition, increases availability of plant nutrients, accelerates decomposition of organic matter, and reduces soil toxicity and soil compaction.

5.5 Rain-fed farming

Crop production that is entirely dependent on rainfall is called dryland agriculture. Depending upon the quantum of rainfall, the dryland agriculture is classified as dry farming, dryland farming and rainfed farming. Crop production without irrigation in arid and desert areas with 40 cm or less annual rainfall is called dry farming. Crop production in semi-arid and sub-humid regions with annual rainfall of 40-120 cm is called dryland farming. Rainfed farming is practiced in humid regions where annual rainfall is more than 120 cm. While irrigation is the primer of modern farming, rainfed farming holds the key for sustainable agriculture of majority of farmers. Of the 142 million hectares of cultivated land in India, 95 million ha is rainfed. In dry land agriculture, scarcity of water is the main problem. Apart from the low and erratic behaviour of rainfall, high evaporative demand and limited water holding capacity of the soil constitute the major constraint in the crop production in dry land area.

Dryland technologies

Crop planning: Selecting suitable crops and varieties capable of maturing within actual rainfall periods will not only help in enhancing production of a single crop but in intensifying the cropping intensity. Under dry land agriculture determination of length of growing period (LGP) i.e., moisture availability of a given soil type, provides better index than total rainfall based crop planning. The LGP is computed as the sum of the period when P is more than 0.5 PET plus time taken to utilize stored soil moisture (assured 100 mm) after P falls short of PET. Under dryland condition such crops should be selected which have capacity to produce good yield under limited soil moisture condition. Usually the crop duration should not exceed the total number of rainy days.

Instead of growing water guzzling crops like rice, non-rice crops can be growing successfully in drylands. Crop substitution has been worked out for different regions which may be adhered to. Even ideal varieties of different crops with drought tolerance have been evolved and these can be safely cultivated in moisture deficit areas.

Instead of growing sole crops, mixed cropping and inter cropping systems need to be adopted in dryland areas to increase cropping intensity. Such cropping can be treated as an insurance crop.

Nutrient management : Soils of dry lands in the country are not only thirsty but hungry also because these soils are severely eroded horizontally as well as vertically. It is necessary to apply all the three major nutrients in adequate amounts. Since soil moisture is limiting in dry lands, the availability of nutrients becomes limited, attempt should always be made to apply fertilizers in furrows below the seed.

Judicious mixing of organic and inorganic fertiliser would be desirable in dryland areas. Organics manures have low nutrient content, but they help improve the moisture holding capacity of soils. In addition to yield advantage, nutrients like potassium help to increase drought tolerance by affecting plant-soil relationship. Transpiration losses are reduced and productivity per unit water increases.

Rain water management: Efficient management of rain water can boost agricultural production from dry lands. The broad bed and furrow system of the International Crop Research Institute for the Semi-Arid Tropics (ICRISAT) for managing rain water in vertisols made it possible to increase crop yields four to five times as compared to normal practice. However, this method could not be adopted widely by the farmers in India because it is costly and labour intensive. The vertical mulching developed at Bellary centres increases the infiltration of water in soil profiles and improves in situ moisture conservation. Ridge and furrow techniques can be adopted in alfisols. Application of compost and farm yard manure as well as raising legumes will add the organic matter to the soil and increase the water holding capacity.

Watershed approach: Watershed management is a holistic approach arrived at optimizing the use of land, water and vegetation in an area and thus, providing solution to alleviate drought. Watershed treatment can prevent soil erosion, improve water availability and increase fuel, fodder and agricultural production on a sustained basis. The major activities in watershed management are (i) improvement of water resources, (ii) In-situ soil and water conservation: rain water harvesting for safe disposal of surface runoff, (iii) increase in cropping intensity, and (iv) alternate land use system for efficient use of lands as per land capability to provide stability in crop productivity.

Alternate land use system: Since all dry lands are not suitable for crop production, alternate land use by taking tree farming, ley farming, dry land horticulture, agro-forestry systems including alley cropping can be successfully followed. This system not only helps in generating much needed off-season employment in mono crop dry land but also minimizes risk, utilizes off-season rains which may otherwise go waste as runoff This would prevent degradation of soils and restore balance in the ecosystem.

5.6 The water-energy-food security system

The water, energy and food security system means that water security, energy security and food security are very much linked to one another. These three sectors are necessary for the benefit of human beings, poverty reduction and sustainable development.

Water-food-energy links lie at the heart of sustainable, economic and environmental development and protection. The demand for all three resources continues to grow for various reasons: a growing population, on-going population movements from farms/villages to cities, rising demand for energy due to higher income of people and climate change.

Water being a finite resource with abundance among the three is the most exploited in food supply chain. In fact, agriculture is the largest user of freshwater, making it responsible for 70% of total global withdrawal, while more than one fourth of energy used worldwide is an input for food production, distribution, and use. In addition, food production and supply chain simultaneously utilize approximately 30% of the total energy that is used globally.

5.7 Key challenges

5.7.1 Water and environment

Nature provides us with multiple benefits of food, fodder, fuel, air, water etc. Water and sediment regimes within natural ecosystems are major factors in determining their health and sustainability. Floods and droughts can have a substantial impact on the ecosystems of wetlands and forests. To ensure that there will be sufficient water to feed a growing and wealthier population, to sustain vital life support systems, and to produce and distribute enough other goods and services, it will be essential to achieve "more crop per drop," "more jobs per drop," "better environment per drop," "improved nutrition per drop," among many such goals.

5.7.2 Agriculture & food production

Higher pressures on water for food production may be expected to develop because large segments of the populations in many countries will tend to raise their standards of living. As salt water constitutes 97 per cent of all global water resources, methods to use sea water for agriculture, which uses 70% of the world's available freshwater resources, would represent a major technological breakthrough that would reduce the freshwater burden. Climate resilient technologies must be adopted to address the impact of climate change. A wide array of "onfarm" agricultural management technologies and practices are available for improving water use efficiency. Change in land use and crop types, improving irrigation efficiency, diversified and intense cropping systems, limiting food waste, water harvesting, supplemental irrigation and precision farming and nutrient management are various strategies for enhancing food production. Further, innovative technologies and investments are required for education and training in the management of water for both irrigated and rainfed farming so as to achieve more productive use of water in agriculture.

5.7.3 Energy production

Humans and their economies and societies critically depend on reliable supplies of energy. Water is also required for production of energy. Efforts to improve water security through construction of desalination and/or wastewater treatment infrastructure and/or large inter-basin transfer projects have significant energy implications. Sustainable energy and water development and production in the future will require an understanding of the interdependencies of both components of such systems. A balance among the needs of all users of energy and water and an appreciation of the impacts on climate, the environment, and the economy will also be necessary. This will likely require some changes in human behaviour.

5.7.4 Water security strategy

Water scarcity can be managed by increasing surface and groundwater storage and allocation through the creation of new infrastructure, desalination of saltwater or brackish water, reuse of wastewater, or recharging aquifers. Reducing water demand can also be achieved by controlling demographic growth, increasing the efficiency of the use of goods that consume water (in particular food products) in their production processes and supply chains, promoting appropriate land use planning, or attenuating the effects of climate change on water through adequate mitigation and adaptation measures. Achieving effective water governance involves implementation of integrated water resources management (IWRM).

5.7.5 More crop per drop

Water management involves integrated management of surface water resources, ground water resources and rainfall. The process of agriculture water management includes the intake, conveyance, regulation, measurement, application of water at right time, right quantity and right method for increased crop production, and effective drainage of excess water that can be detrimental to soil environment.

Water and food security: Food security is required to be achieved at household level and national level. Food security has four dimensions such as availability, access, absorption and stability. Water is very much linked with food security. There should be enough production to make food available to all as per requirement. There should be more crops per drop of water. In India out of 142 m ha of net cultivated area nearly 53 m ha (37%) is irrigated and 89 m ha (63%) is rainfed. Irrigated agriculture contributes to 55% of production whereas 63% of cultivated area under rainfed situation produces only 45% of total production. Thus water, either received from irrigation or rainfall, is very important for food security. It is necessary that the farmers should avoid much of water guzzling crops and increase water use efficiency in order to achieve food security. The farmers should produce more crops per drop.

Water availability and rainfall in India are dependent on two monsoons, the south-west (summer) and north-east (winter). Most rainfall occurs between the months of June and September, with the average annual rainfall approximately 1170mm. There is, however, considerable variation between regions. The current average per capita availability of water is 1,600m^3 per year. Population forecasts indicate that by 2050 this average will be reduced to approximately 1,000m^3 per year. The overexploitation of groundwater, a lack of storage capacity and increasing levels of pollution in a business-as-usual scenario, will greatly increase the risk of severe water insecurity across India.

Virtual water: Virtual water is the amount of water required to produce a product, from start to finish which is normally divided into three categories for classification: blue, green and grey. Blue water is the water contained in rivers and lakes and is the water processed by the water companies to supply public and commercial demand; this water is used in the food processing industry. Green water is the water supplied through rainfall and contained in soils; the majority of agricultural production is based on this water. Traditionally grey water is wastewater. Blue and green are two types of natural water. Blue water exists in flows and storages of freshwater that exist at the land surface and in groundwater systems. Blue water can be pumped and used for various purposes. Water required for production of some products is indicated in Table 5.5 .

Table. 5.5. Water required for production of some items (virtual water)

Item	Consumption of water (Litre) for production of one kg.	Item	Consumption of r wate in Litre for 1 kg
Wheat	1200	Egg	2700
Rice	3000-4000	Milk	790
Vegetables	5460	Steel	150
Potato	262	Paper	900
Mango	1600	Synthetic rayon	2000
Banana	272	Beef	13500

Quality of water

Both irrigation water quality and proper irrigation management are critical to successful crop production. The quality of the irrigation water may affect both crop yields and soil physical conditions, even if all other conditions and cultural practices are favourable/ optimal. There are three principal problems that can arise from the quality of irrigation water delivered to the agricultural fields. These are salinity, sodicity and toxicity. Even heavy metal in water can contaminate the crops and become harmful to human and animal health. The parameters which determine the irrigation water quality are divided to three categories: chemical, physical and biological. High level of salts in the irrigation water reduces water availability to the crop (because of osmotic pressure) and causes yield reduction. High sodium absorption ratio (SAR) levels might result in a breakdown of soil structure and water infiltration problems. The quality of the irrigation water can be also determined by toxicity of specific ions. Toxicity occurs within the plant itself, as a result of accumulation of a specific ion in the leaves. The most common ions which might cause a toxicity problem are chloride, sodium and boron. Special attention should be given to boron because toxicity occurs in very low concentrations, even though it is an essential plant nutrient.

Toxic levels can be managed by proper leaching, increasing the frequency of irrigations, avoiding overhead irrigation, avoiding the use of fertilizers containing chloride or boron, selecting the right crops, etc.

References

Abdulmumin, S., Bastiansen, J., Smith, M., Rijks, D. and Gbeckor-Kore N.A. 1990. Application of climatic data for irrigation planning and management. Training Manual. FAO WMO. Rome.

Attila Yazar, 2013, On-farm agricultural water management (irrigation scheduling techniques)

Brady, N. and R. Weil. 2002. The Nature and Properties of Soils, 13th Edition. Prentice Hall. Upper Saddle River, New Jersey. 960 p.

Cassel, D. K. 1984. Irrigation scheduling. Crops and Soil Magazine, Am. Soc. Agron. February, March-April.

Jones, J.W .,2004. Irrigation scheduling: advantages and pitfalls of plant-based methods. Journal of Experimental Botany, Vol. 55, No. 407, Water-Saving Agriculture Special Issue, pp. 2427–2436, November 2004

Martin, D.L., Stegman, E.C., and Fereres, E., 1990. Irrigation Scheduling Principles, In:. G.J. Hoffman et al.(eds), Management of Farm Irrigation Systems. ASAE Monograph. St. Joseph, MI: 155-1990

Parihar, S. S., Gajri, P. R. and Narang, R. S. (1974). Scheduling irrigation to wheat using pan evaporation. Indian J. Agric. Sci., 44 (9) : 567-571

Penman, H.L.1948, Natural evaporation from open water, bare soil and grass, *Proc, Roy, Soc.*Lond, 193:120-45

Penman, H.L.1956. Evaporation: An introductory survey, *Neth, J.agric.Sci*, 4:9-29

Pereira A.R., Green S.R., Villa Nova N.A. 2006. Penman–Monteith reference evapotranspiration adapted to estimate irrigated tree transpiration. Agric Water Manag 83:153–161

Perrier, E.R., and Salkini, A.B. 1987. (eds.) Supplemental lrrigation in the Near East and North Africa. Kluwer Academic Publishers, London

Prihar, S.S, Sandhu, B.S.1987.Irrigation of Field crops: principles and Practices (ICAR)

Wanjura, D.F., Upchurch, D.R. and Mahan, J.R. 1990. Evaluating decision criteria for irrigation scheduling of cotton. Transactions of the ASAE. 33 (2):512-518

Yazar, A., Kanber R, Ozekici. B., 1995. Irrigation scheduling in agronomic practice. In Sustainability of Irrigated Agriculture (edts) L.S. Pereira et al. NATO ASI Series E: Applied Sciences-Vol. 312: 251-266

6

Integrated Pest Management

6.1 What is a pest?

Insects and mites that damage crops, pathogens that cause diseases in plants, weeds that compete with field crops for nutrients and water, plants that choke irrigation channels or drainage systems, rodents that eat young plants and grain, and birds that eat seedlings or stored foodstuffs are all crop pests. As plants were domesticated and grown, increasingly in monoculture, the potential damage caused by their co-evolved pests and disease pathogens were realised. A pest can be defined as anything that people consider a threat to themselves, their crops, animals or property. The crop pests include insects, weeds, nematodes, rodents, molluscs, mites, fungi, bacteria, virus, phytoplasma and viroid. The ravages of pests on crops have been recorded since the earliest times in civilisation.

6.2 Pest control measures

6.2.1 Indigenous methods

Indigenous methods of pest control were limited to cultural and mechanical measures along with use of some locally available materials to manage the pests.

i) **Cultural practices:** Ploughing, hoeing and basin preparation expose soil inhabiting insect, pests and other arthropods and nematodes to harsh weather and to natural predators. Insects are most vulnerable when in the pupal stage and most insect-pests pupate in the soil which furnishes a protective habitat. Birds pick up the exposed pupae following these cultural operations. Some insects e.g. grasshoppers, crickets, mole-crickets and borers lay eggs in the upper layers of the soil which are exposed during ploughing and killed. Some insects like cutworms, grubs of the root borer and white grubs which feed on the root system of plants are also exposed after ploughing and killed. Deep ploughing in winter helps reducing the overwintering populations of several pests. Besides dislodging the pests from their protective habitat and subjecting them to unfavourable conditions for survival, these scientifically tempered cultural practices also improve aeration of the soil and facilitate proper percolation of water into the soil.

ii) **Mechanical measures:** "Catch and kill" was the most popular method of pest control. Hand picking of pests and their destruction is a time tested method of pest control. The collected pests are destroyed by immersing them in kerosinized water or by deep burying. Nocturnal insects responding positively to light, e.g. defoliating beetles, moths of Bihar hairy caterpillar, tomato fruit borer, tobacco caterpillar etc. are collected, using light traps and are subsequently destroyed.

iii) **Cow-dung and clay mixture:** The practice of applying a thin paste made of cow dung, clay and cow urine, as a disinfectant, on the floor of mud (*kutcha*) houses in rural areas is an age old practice. Similarly, the application of a paste, of good consistency, to treat wounds and injured limbs of fruit trees has been in vogue, in villages, since long. Such a paste is also applied to pruned twigs/limbs. A fine slurry is prepared by thorough mixing of the clay, cow dung and cow urine in a container. The paste is then applied manually using bare hands or a locally made brush. Cow dung and cow urine possess complex degrading substances and may possess anti- bacterial properties. Sealing of wounds/cut by cow dung or clay prevents access of pathogens.

iv) **Pruning of fruit trees:** Pruning and training of fruit trees is an important practice performed during the dormant season to create a proper frame and provide symmetry to the tree for a balanced growth. Pruning of old, damaged and weak portion of the tree encourages new growth which is healthier and stronger. Thus, pruning aids pest control as it releases overwintering population, improves general health of the tree and helps in maintaining a balanced foliage distribution with proper aeration and sunlight penetration.

v) **Use of wood ash on and around vegetable crops:** It is a common practice to sprinkle wood ash on vegetable crops, especially growing in kitchen gardens. A thick layer of ash is spread on the soil around plants and on foliage to protect against a variety of pests. This acts as a physical poison usually causing abrasion of epicuticular waxes and thus exposing pests to death through desiccation. The treated foliage becomes unpalatable for foliage feeders like cutworms, caterpillars, grasshoppers, etc. Ash also provides potash to the plants.

vi) **Kerosene oil for killing borers:** Farmers have been using kerosene oil to kill tissue borers using a flexible metallic wire which is inserted through the hole made by the borer into the gallery to clean it. Then a small bung made of cloth soaked in kerosene oil is inserted into the hole and finally, it is plugged using a paste of cow dung and clay or even by cement. The inserted metallic wire into the gallery causes physical injury to the larvae which are destroyed in the process..

Use of Indigenous technical knowledge (ITK): There is a wealth of ITKs in rural and tribal areas where many cultural, mechanical, botanical methods are used for pest control. Slash and burn system (Shifting cultivation) was practiced in hilly areas for growing many crops in mixed pattern which also provided a pest control measure. Many botanicals prepared from Neem, Karanj, tobacco, garlic, chilli etc., were used for pest control. The ethnic farming communities are the store house of indigenous knowledge of technical knows how regarding the overall management of indigenous crops. Cultural practices such as mixed / multiple cropping, zero tillage, clean cultivation, slash and burning , green manuring, sequential cropping and harvesting, fallowing, flooding etc. were used for control of crop pests.

6.2.2 Conventional pest control

The conventional system of pest control focused much more on the elimination of pests solely through chemical means like pesticides. In chemical control little or no attention was paid to the potential damage done by pesticides on the environment or to their effects on the food and water of either humans or animals. Indiscriminate use of pesticides that results killing of natural enemies of pests, beneficial soil microorganisms and its residual effect in food chain necessitates the development of integrated pest management concept to minimize and judicious application of pesticides as last resort if other eco-friendly alternative methods fail to contain the harmful agricultural pests.

These chemicals were so successful that research into alternative methods of pest control was often downgraded or abandoned. This had a marked effect on areas of research such as biological control. The publication of the book "Silent Spring" by Rachel Carson in 1962, and the subsequent reports of widespread pesticide resistance, has served to turn attention to the use of biological controls in recent times. Carson's book led to the withdrawal of DDT and a global ban on the use of the chemical. Biological control also has a long history as a component of pest management in crops.

Another Book **"Beyond Silent Spring"** written by van Emden, H.F., Peakall, David B. (1996) after 32 years outlines a new paradigm, "Agro-ecological Intensification of Crop Protection", which reduces negative impacts on the environment and enhances the provision of ecosystem services. It discusses the use of ecologically based management strategies to increase the sustainability of agricultural production while reducing off-site consequences, highlighting the underlying principles and outlining some of the key management practices and technologies required implementing agro-ecological pest management. It also comprehensively explores important topics like stimulo-deterrent diversion strategy, precision agriculture, plant breeding, nutrient management, habitat management, cultural approaches, cultivar mixtures/multiline cultivars, crop rotation, crop residue management, crop diversity, cover crops, conservation tillage, bio-fumigation, agro-forestry, and addition of organic matter.

6.2.3 Ecological pest management

Each grower has their own strategy for producing crops, minimizing crop losses, and making a profit in a manner that is acceptable to the retailer, safe for the consumers, and less disruptive to the environment. In any agro-ecosystem there are biotic and abiotic components (Fig. 6.1). In biotic component there are producers, consumers and decomposers. There is a natural balance maintained among the abiotic factors, producers and consumers. While plants are the producers, humans, animals, birds, crop insect pests are primary consumers. Crop pests take out certain shares of plant and cause crop loss. There are also another sets of organisms, called defenders or natural enemies which control the crop pests by predating, parasitizing or in other ways so that a delicate natural balance is maintained (Fig. 6.2).

Components of an Ecosystem

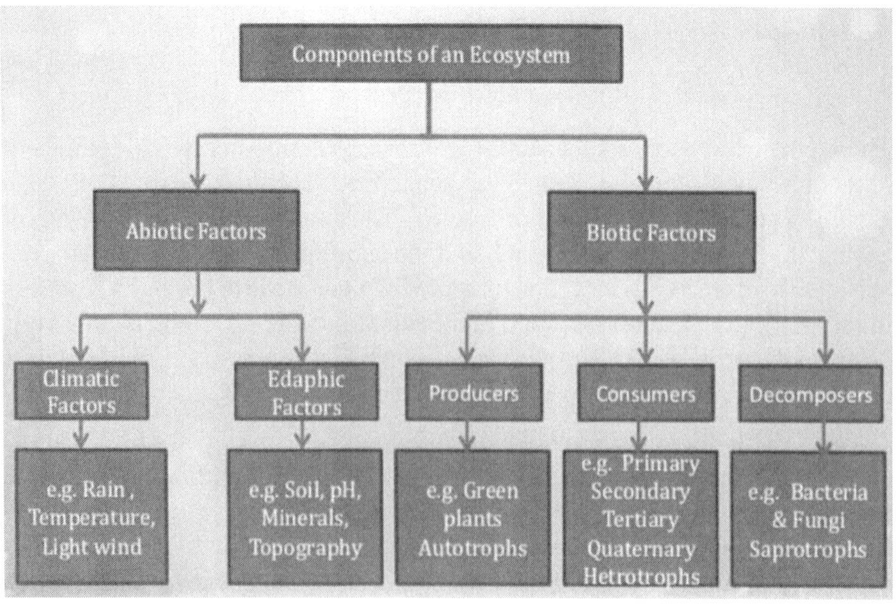

Fig. 6.1. Components of ecosystem

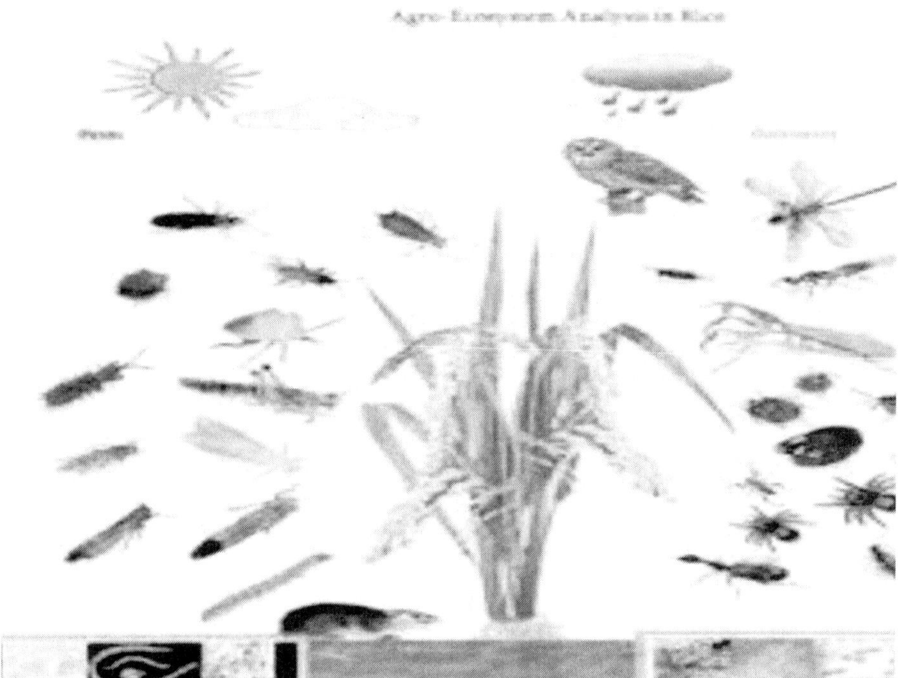

Fig. 6.2. Crop pests and natural defenders

When there are certain environmental problems or human activities that kill the population of natural defenders, the pest outbreak becomes a problem. Therefore the ecological aspects of pest control will be required to maintain the natural balance in the agro-ecosystem.

Economics and yield loss relationship

Economic threshold: The criterion most often used to determine the amount of damage (or level of pest population) which can be tolerated is the economic threshold (Stern et al., 1959). Above this threshold the cost of control of the pest is out-weighed by the potential losses due to the pest. Carlson and Main (1976) considered economic thresholds to be based upon the cost of pest control, the value of crop loss which could have been prevented and the forecast of pest and disease intensity so that timely of pest control can be determined. The compensation point, or tolerance level, is dependent on the growth stage of the crop when the pest becomes evident, as well as crop management practices, geographic location and climatic factors (Main, 1977).

Economic Damage (ED): The ED begins at the point at which the cost of crop damage equals the cost of control

Economic Injury Level (EIL): The lowest pest population density that will cause ED

Economic Threshold (ET) or Action Threshold (AT): The population density at which control action should be determined (initiated) to prevent an increasing pest population (injury) from reaching the EIL. To make a control practice profitable, or at least break even, it is necessary to set ET below EIL.

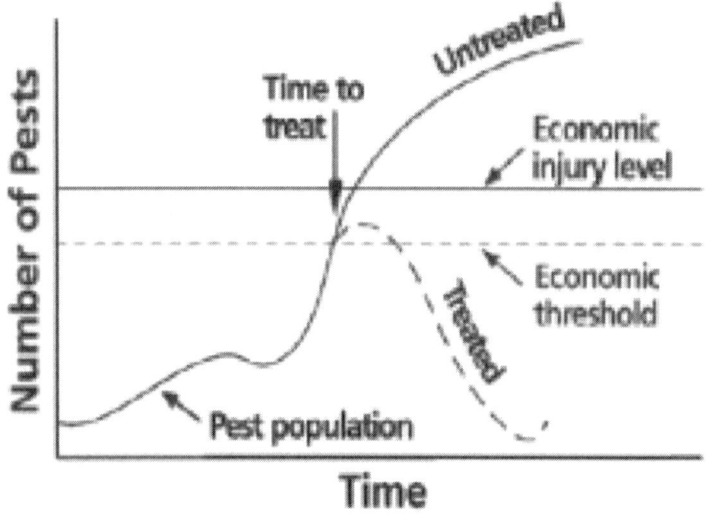

Fig. 6.3. EIL and ETL of pests

The environment and pest

The environment can also have a direct effect on insects and diseases and their interactions with their host. Temperature is the most important influencing factor in insect development. The amount of heat required (physiological time) can be used to predict the time required for an insect to complete a developmental process or stage. This physiological time is calculated from the summation or accumulation of time by temperature and is expressed as day degrees (Southwood, 1978). The transmission of viruses can be affected by the mobility of vectors, which is affected by temperature and the interaction between the virus and its vector (non-persistent or persistent) (Jeger et al., 1998; Irwin 1999). Temperature can also be considered one of the most important direct determinants of disease progress. In most cases, the optimum temperature for disease development mirrors the optimum for the host and the pathogen. Moisture (rain, dew, and irrigation) is also crucial to infection by most plant pathogens and plays a role in dispersal of many such pathogens (West et al., 2001).

6.3 Integrated pest management (IPM)

The concept of pest management aims at a balanced approach to managing pest populations to levels that do not cause economic losses rather than eliminating or eradicating them completely. IPM suggests a basket of options and the farmers can choose the appropriate measure for their situations without adopting blanket recommendations or prescriptions.

Integrated Pest Management is a broad ecological approach to manage the crop pests in an ecologically viable, socially acceptable and environmentally safe manner. It aims at managing the crop pests within economic injury level. This essentially is IPM It includes a variety of options, any one of which may not significantly reduce the pest population, but the sum total of which will give adequate reduction to prevent economic losses.

Integrated Pest Management (IPM) is a philosophy that involves the management of a pest instead of controlling or eradicating a pest. It requires a greater knowledge of the pest, crop and the environment. Therefore, its strategy focuses on harnessing inherent strengths within ecosystems and directing the pest populations into acceptable bounds rather than toward eliminating them. This strategy avoids undesirable short term and long term ripple effects and will ensure a sustainable future (Lewis et al., 1997). Integrated pest management is a comprehensive long term pest management program based on knowledge of an ecosystem that weighs economic, environmental, and social consequences of interventions (Flint and van den Bosch, 1981).

The word 'integrated' in IPM initially referred to the simultaneous use or integration of any number of tactics in combination, with focus on maintaining a single pest species below its economic injury level. Although, in theory a single strategy results from the simultaneous integration of several tactics, in practice, the integration actually occurs in a step-wise, time delayed fashion. Several of the tactics are compatible, but some are not. Certainly the tactics of biological control, habitat manipulation, and legal control go alongside. The tactic of host resistance can stand alone or be combined with the other tactics just mentioned. Chemical control is generally compatible with host resistance. Thus, a management strategy integrates one or several compatible tactics into a single package (Irwin, 1999).

Principles of control

Usually preventive and curative controls are adopted in pest control programme. Preventive control methods are used to reduce the number of contacts, or the effect of those contacts between pest and host. Curative control method is a deliberate attempt for the destruction of pests. The principles of exclusion, avoidance, protection and eradication, applied either as regulatory, physical, cultural, biological or chemical methods of control are adopted in both the control measures.

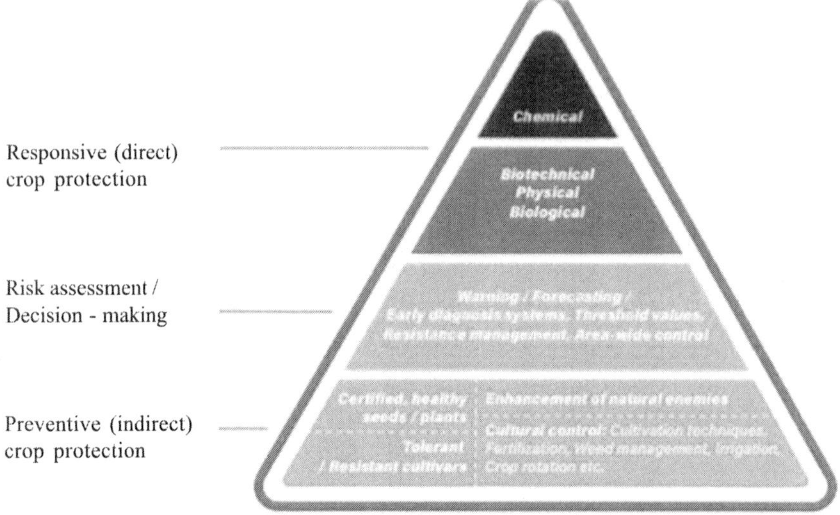

Fig. 6.4. Principles of IPM

General crop management practices should aim at increasing crop vigour to resist pest attack and create an environment unfavourable for the insect pests and diseases to grow. All such management practices are implied in the principle of crop protection. Biological, physical, chemical and cultural control methods all have a role to play. Field sanitation is also important to keep the crops free from pest attack. The destruction of the pest from crop is the function of eradication techniques.

6.3.1 Ecological principles

There are certain ecological principles to manage the crop pests. It is well established-and known by most farmers-that crop rotation can decrease many disease, insects, nematodes, and weed pressures. A few other examples of management practices that reduce pest pressure are as follow:

- Insect damage can be reduced by avoiding excess inorganic nitrogen levels in soils by using better nitrogen management.
- Adequate nutrient levels reduce disease incidence.
- Damage from insect and disease (such as fungal diseases of roots) can be decreased by lessening soil compaction.

- Severity of root rots and leaf diseases can be reduced with composts that contain low levels of available nitrogen but still have some active organic matter.
- Many pests are kept under control by having to compete for resources or by direct antagonism from other insects (including the beneficial feeding on them). Good quantities of a variety of organic materials help maintain a diverse group of soil organisms.
- Root surfaces are protected from fungal and nematode attack by high rates of beneficial mycorrhizal fungi. Most cover crops help keep mycorrhizal fungi spore counts high and promote higher rates of infection by the beneficial fungi.
- Parasitic nematodes can be suppressed by selected cover crops.
- Weed seed numbers are reduced in soils that have a lot of biological activity, with both microorganisms and insects helping the process.
- Weed seed predation by ground beetles is encouraged by reduced tillage and maintenance of surface residues.
- Residues of some cover crops, such as winter rye, produce chemicals that reduce weed seed germination.
- Use of organisms like predators, parasites, parasitoids, or pathogens that attack harmful insect, weed or plant disease or any other organism in its vicinity. Many soil microorganisms are antagonistic. They secrete a potent enzyme which destroys other cells by digesting their cell walls and degrade the cellular material as well as released protoplasmic material.

Ecological principles provide a good framework for sustainable management, but we must also recognize that crop production is inherently an "unnatural" process because we favour one organism (the crop plant) over the competing interests of others. With currently available pesticides, the temptation exists to simply wipe out competitors, for example through soil fumigation, but this creates dependency on purchased materials from off the farm and weakens the overall resiliency of the soil and cropping system. The goal of ecological crop and soil management is to minimize the extent of reactive management (which reacts to unanticipated occurrences) by creating conditions that help to grow healthy plants, promote beneficial organisms and stress pests.

6.3.2 Components of IPM

Five general types of single component control methods may be used in IPM programmes in stored ecosystems.

Fig. 6.5. IPM measures

6.3.3 Management options

Host plant resistance: Host plant resistance involves breeding varieties with desirable economic traits, but less attractive for pests, for egg laying and subsequent development of insect, disease or nematode. It also involves withstanding the infestation/infection or the reduction of pests to level that they are not large numbers during the plant growth period (Sharma, 1997).

Physical or mechanical control : Sun drying, mechanical sieving of seeds, temperature treatment (high or low), heat treatment of soil, soil sterilisation, refrigeration are some of the physical and mechanical measures of pest control. Use of traps (light trap, sticky trap, pheromone trap, yellow trap), bamboo perchers etc. are important mechanical measures.

Mechanical sieving and grading of seed and separation by flotation are important means by which seed for sowing can be freed of various pathogen propagules such as galls and sclerotes. Tillage can also help in the control of diseases. Cultural practices such as deep ploughing of diseased crop residue and burning of crop stubble are important physical means used to suppress disease caused by Sclerotinia spp. and other organisms which survive in stubble, but which produce aerial spores to infect subsequent and adjacent crop. Refrigeration is probably the most widely used method of controlling post-harvest diseases of fleshy products (Agrios, 1997). The reduced temperatures do not kill the pest or pathogen, but reduce their growth or activity. Gamma irradiation has also been found to be a satisfactory method of destroying pests in fruits such as tomatoes and strawberries (Agrios, 1997).

Cultural control: Cultural control techniques involve practices that modify the environment to favour crop growth but discourage or avoid conditions that favour pest development. Cultural methods used to control pests and disease, include crop rotation, mineral nutrition, time of sowing, special cultivation, water management, crop rotation, sanitation and clean seed. Sanitation includes all activities aimed at eliminating or reducing pathogen levels in association with plants, paddocks, machinery and storage facilities.

Crop seed is a natural repository for a wide spectrum of pests (insects, weeds and diseases) and hence is an ideal vehicle for the introduction and spread of new pests and diseases into 'clean' areas. Seed-borne diseases due to causing organisms (viruses, fungi, bacteria, nematodes) can be disseminated by seed. Hence use of clean seed and seed treatment before sowing are very important.

The use of crop varieties resistant or tolerant to particular diseases is one of the main methods of disease control. Host plant resistance is the inherent ability of crop plants to restrict, reduce or overcome a pest infestation (Kumar, 1984). This definition includes the concept of tolerance, in which a plant variety will be able to tolerate a level of pest infection without a reduction in yield. Many authors consider that tolerance should be treated separately from true resistance (Keane and Brown, 1997; Beck, 1965).

Modifying irrigation practices, fertilizer application, and other agronomic practices can create conditions that are less suitable for the pest and favourable for the natural enemies. Destroying crop residue and thorough cultivation during summer or after harvest of a crop will eliminate breeding sites and control soil-inhabiting stages of the pest. Sanitation practices to remove infected/infested plant material, regular cleaning field equipment, cleaning of field bunds, avoiding accidental contamination of healthy fields through human activity are also important to prevent the pest spread. Crop rotation with non-host or tolerant crops will break the pest cycles and reduce their build-up year after year.

Behavioural control

The behaviour of the pest can be exploited for its monitoring and control through baits, traps, and mating disruption techniques .Baits containing poisonous material will attract and kill the pests when distributed in the field or placed in traps. Pests are attracted to certain colours, lights, odours of attractants or pheromones.

Biological control : Biological control can be considered to include control based on the management of some aspect of the biology of the pathogen species - genetic manipulation, breeding for resistance in crop plants, crop rotation, grazing management - aspects which are considered separately. In the more specialised sense, it refers to the manipulation by man, of introduced and indigenous natural enemies of the disease organism/pest in order to suppress it. Plant protectionists, in particular, adopt a more general interpretation which has been modified from Garrett (1965) as 'any condition under which, or practice whereby, survival or activity of a pest is reduced through the agency of any other living organism (except man)'.

The suppression of insects and weeds by the use of natural enemies has seen more success than the use of these types of agents against plant diseases. Bio-control agents of an introduced organism may be selected from within the target country, screened for pathogenicity, formulated and used in a similar way as a biocide for control. This approach lends from both biological control and biocide application and the agent is termed a bio-pesticide.

Biological control may be defined in a narrow sense as "the manipulation of predators or pathogens to manage the density of an insect population". In a broader sense, it includes "the manipulation of other biological facets of the pest life system, such as its reproductive processes (i.e. sterile male technique), its behaviour (pheromones), the quality of its food and so forth."

Natural enemies such as predatory arthropods and parasitic wasps can be very effective in causing significant reductions in pest populations in certain circumstances. Periodical releases of commercially available natural enemies or conserving natural enemy populations by providing refuges or avoiding practices that harm them are some of the common practices to control endemic pests. Some natural enemies in pest control are indicated in Table 6.1.

Table 6.1. Biological control of pests

Pest	Natural enemy
Aphids	Ladybird beetle, Green lacewings, Spiders, Syrphid flies
Beetles	Anaphes species, Damsel bugs, Elm leaf beetle parasite, Spiders, Leather winged beetle
Bugs	Green lacewings, Spiders, Damsel Bugs
Caterpillar	Assassin bugs, Brown lacewings, Damsel bugs, Spiders, Trichogramma spp., Green lacewings
Leaf hopper	Spiders, Damsel Bugs,
Mealy bugs	Brown lace wing, Green lacewing, Dusty wing
Mites	Damsel bugs, Spider mite destroyer lady beetle, Green lacewing
Scales	Brown lacewing, Green lacewing
Thrips	Damsel bug, Spider
White fly	Spider, Dusty wing, Green lacewing, Brown lacewing

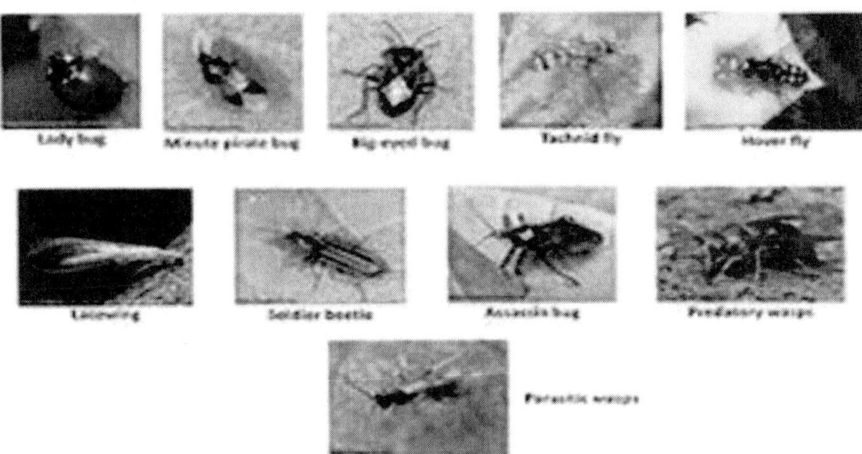

Fig. 6.6. Natural enemies on pest control

Chemical control: Pesticides and their judicial use are essential to modern agriculture. After their introduction in the 1950s, insecticides were viewed as the universal panacea for all pest problems. However, following the revelations in Rachael Carson's book and the development of resistance to many pesticides, a more rational approach to the use of pesticides has developed. This relies on an understanding of pest population dynamics, the mode of action of the pesticide, application technology, safe use and toxicology (Dent, 2000). However, to be technically accurate, chemical control should include synthetic chemicals as well as chemicals of microbial or botanical origin. Although botanical extracts such as azadirachtin (neem) and pyrethrins, and microbe-derived toxic metabolites such as avermectin and spinosad are regarded as biologicals), they are still chemical molecules, similar to synthetic chemicals, and possess many of the human and environmental safety risks as chemical pesticides.

References

Agrios GN (1997) Plant Pathology. 4th edition. Academic Press Ltd., San Diego, California

Beck SD (1965) Resistance of plants to insects. Annual Review of Entomology. 10, 207-232

Carlson GA and Main CE (1976) Economics of disease-loss management. Annual Review of Phytopathology 14, 381-403

Carson, Rachel (1962). *Silent Spring*. Houghton Mifflin Company.

Dent, D (2000) Insect Pest Management. 2nd ed. CABI Publishing, Cambridge

Flint, M. L. and van den Bosch, R. (1981) Introduction to Integrated Pest Management (Plenum, New York).

Garrett SD (1965) Toward biological control of soil-borne plant pathogens. In Baker, K.F. and Synder WC (eds) Ecology of Soil-Borne Plant Pathogens: Prelude to Biological Control. University of California Press, Berkeley, 4-17

H.F. van Emden (Author), David B. Peakall (1966) : Beyond Silent Sporing: Integrated Pest Management and Chemical Safety.

Irwin ME (1999) Implications of movement in developing and deploying integrated pest management strategies. Agricultural and Forest Meteorology 97, 235-248

Irwin, M.E. (1999) Agricultural and Forest Meteorology 97: 235–248

Jeger MJ, Vandenbosch F, Madden LV and Holt J (1998) A model for analysing plant-virus transmission characteristics and epidemic development. Journal of Mathematics Applied in Medicine and Biology 15, 1-18

Keane P and Brown J (1997) Disease management: resistant cultivars. In JF Brown and HJ Ogle (eds.) Plant Pathogens and Plant Diseases. Rockvale Publications, Australia

Kumar R (1984) Insect Pest Control with Special Reference to African Agriculture. Edward Arnold, London

Lewis, W. J., van Lenteren, J. C., Sharad C. Phatak, and Tumlinson, J. H. (1997) Proc. Natl. Acad. Sci. USA 94: 12243–12248

Main CE (1977) Crop destruction - the raison d'etre of plant pathology. In Horsfall, J.G. and Cowling, E.B. (eds) Plant Disease an Advanced Treatise. Volume I. How Disease is Managed. Academic Press Inc., NY pp 55-78

Sharma, B.K. (1997) Environmental Chemistry, Krishna Ltd., Delhi pp. soil-114

Southwood TRE (1978) Ecological Methods, with Particular Reference to the Study of Insect Populations. 2nd ed. Chapman & Hall, London

Stern VM, Smith RF, van den Bosch R and Hagen KS (1959) The integrated control concept. Hilgardia 29, 81-101

West JS, Kharbanda PD, Barbetti MJ and Fitt BDL (2001) Epidemiology and management of Leptosphaeria maculans (phoma stem canker) on oilseed rape in Australia, Canada and Europe. Plant Pathology 50, 10-27

7

Diversified Farming Systems

7.1 Farming system

Farming system is a complex inter-related matrix of soil, plants, animals, implements, power, labour, capital and other inputs controlled in part by farm families and influenced by varying degrees of political, economic, institutional and social forces that operate at many levels. In farming system there is integration of farm enterprises such as cropping systems, animal husbandry, fisheries, forestry, sericulture, poultry etc. for optimal utilization of resources bringing prosperity to the farmer. In farming system, the farm is viewed in a holistic manner. Sustainability is the objective of the farming system where production process is optimized through efficient utilization of inputs without infringing on the quality of environment with which it interacts on one hand and attempt to meet the national goals on the other.

Objectives of the farming system

The broad objectives of farming system are productivity, profitability, sustainability, energy conservation and employment generation. The specific objectives are:

- Increasing productivity and profitability
- Sustainability in production system with environmental protection
- Production of balanced food, fodder, fuel, timber etc. with conservation of energy
- Year round income and employment generation.
- Increase in input use efficiency and providing scope for agro-industries.

Principles

- Minimization of risk
- Recycling of wastes and residues
- Integration of two or more enterprises
- Optimum utilization of all resources

- Maximum productivity and profitability
- Ecological balance
- Generation of employment potential
- Increased input use efficiency

7.2 Diversified farming system

The ultimate goal of sustainable agriculture is to develop farming systems that are productive and profitable and can conserve the natural resource base and protect the environment sustainably. The means of achieving this is low input methods and skilled management, which seek to optimize the management and use of internal production inputs (i.e., on-farm resources) in ways that provide acceptable levels of sustainable crop yields and livestock production and result in economically profitable returns.

Diversification in agriculture helps reduce risk and increases profitability in farming. It is seen that diversification in high value cash crops can possibly increase farm income. Change in consumer taste and preference also lead to increase demand of various value added product which come from agriculture. Agriculture is producing feed, fibre, fuel, food and other goods by systematic growing and harvesting of plants and animals (De and Jirli, 2017). In the overall perspective, with the introduction and adoption of modern technologies, agriculture sector is expected to achieve a vertical growth (Himani B. Patel, 2016). Diversified farming systems allow critical ecosystem services – like pollination and pest control – to be generated and regenerated within the agro-ecosystem, aided by the human knowledge to sustain those processes. It is critical to feed the world population reliably and in perpetuity while mitigating climate change and avoiding a collapse of the ecological systems on which human survival depends. DFS builds soil fertility, ensures cycling nutrients and water, and supports beneficial insects that control pests and pollinate crops. These practices include planting many crop varieties in a single field; incorporating trees, livestock, or aquaculture; rotating crops; planting hedgerows and riparian buffers; and conserving natural areas in the landscapes around the farm.

Advantages

- Better use of land, labour and capital through adoption of crop rotations, steady employment of farm and family labour and more profitable use of equipment are obtained in diversified farming.
- The farmer and labour engaged all the year round in different activities.

- Less risk to crop failure and market price of the product.
- The by-products of this farm can utilize properly as cattle, poultry, birds, etc. are reared with crop production.
- Regular and quicker return is obtained from various enterprises.
- Soil erosion can be checked as land kept under cultivated throughout the year
- Soil fertility can be checked as land kept under cultivated throughout the year.
- Diversified farming is less risky than specialized farming
- Best use of all equipment and farm machinery.

7.3 Cropping systems

In an agricultural production system cropping patterns and cropping systems are usually discussed. A cropping system refers to the principles and practices of cropping and their interaction with farm resources, technology, aerial and edaphic environment to suit the regional or national or global needs and production strategy. It is an important component of farming system. A cropping system is a community of crops managed in a farm unit to attain various human goals such as food, fibre, feed, raw materials, wealth and satisfaction. When cropping system is integrated with livestock, poultry, fishery, forestry etc. it is called farming system.

7.3.1 Cropping pattern

A cropping pattern refers to annual sequence and spatial arrangement of crops and fallow on a given land.

Monoculture is a practice of growing same crop year after year in the same land. This type of cropping is not desirable for sustainable agriculture. On the other hand, polyculture is a system of growing two or more crops simultaneously with or without animal component. When only different crops are involved it is called multiple cropping and when crops and animals are involved it is called mixed farming.

7.3.2 Multiple cropping

Multiple cropping is a way to grow more food on the same piece of land. Multiple cropping refers to intensification of cropping both in time and space. It includes sequential cropping, inter-cropping and mixed cropping. It is defined as growing multiple crops sequentially or as intercrops, within a year. It increases crop diversity, increases production and profitability. The choice of species for multi-cropping varies with geographical location, level of input, and local preference.

Multiple cropping is probably the oldest form of agriculture and is still widespread, particularly in tropical and low-input production systems. In principle, sequential cropping, where a second crop is grown after the harvest of a first crop, is a simple form of multi-cropping. However, most multi-crops involve the cultivation of two or more species on the same piece of land where the growth cycles of different species overlap for at least part of their duration. Such multi-crops may take the form of intercrops, where growth and yield are confined to a single growing season, or agroforestry, where a perennial component, e.g., trees or shrubs, occupies some of the land over a series of growing seasons. These complex systems may include dual canopy crops where high-value plantation crops such as oil palm, coconut, cocoa, or rubber are inter-planted with annual species.

Different types of cropping systems

a) **Sequential cropping**: Growing two or more crops in a sequence on the same field in a cropping year is called sequential cropping. The succeeding crop is planted after the preceding crop has been harvested under irrigated situation or with residual soil moisture. There is no inter-crop competition. Farmers manage only one crop at a time in the same field.

b) **Intercropping**: It refers to growing of two or more dissimilar crops simultaneously on the same piece of land. Here a base crop is grown with a companion crop. The recommended optimum plant population of the base crop is suitably combined with appropriate additional plant density of the associated/component crop. The objective is intensification of cropping both in time and space dimensions and to raise productivity per unit area and inputs by increasing the pressure of plant population. There are four types of inter-cropping systems.

Mixed inter-cropping: Growing of component crops simultaneously with no distinct row arrangement in this system. This is commonly used in labour intensive subsistence farming situations. Mixed cropping is an old system of cropping where cereals are mixed with subsidiary crops like pulses for an assured production system and maintenance of soil fertility. In many cases crop mixtures generally give better results than monoculture under dryland conditions. In shifting cultivation or *jhooming* 5-10 crops like rice, millets, pulses and oilseeds of different maturity durations are cultivated after slash and burn, which ensures household food security from September to May and June.

Row inter-cropping: In this system growing component crops simultaneously in different row arrangement are taken up. This is used in mechanized agriculture.

Agronomists have taken up intercropping in place of mixed cropping with different row combinations of main and companion crops. For example, eight rows of rice are planted between two rows of arhar. Many such crop combinations with crop geometry have been experimented in Dryland Research Stations like Phulbani in Odisha. The land equivalent ratio (LER) of mixed or intercropping has always been higher than the mono-cropping. So it is a vertical rise of crop production. One popular traditional example of intercropping is planting corn, beans, and pumpkin together. Corn plants provide support to beans, which fixes atmospheric nitrogen and supplies to soil that corn also uses.

Strip inter-cropping: Component crops are grown in different strips wide enough to permit independent cultivation but narrow enough to the crop to interact agronomically.

Relay inter-cropping: Component crops are grown in relay, so that growth cycles overlap. It refers to planting of succeeding crop before the harvest of preceding crop and planting of succeeding crop may be done before or after attainment of reproductive stage, completion of active life cycle, senescence of leaves or attainment of physiological maturity. Examples include paddy rice and legumes, cereals and legumes, cereals and other cereals.

c) **Agro-forestry**: It is a land use systems in which woody perennials (trees, shrubs) are grown in association with herbaceous plants (annual or perennial crops, or pastures) and/or livestock. In this system the woody component/s may provide more than one product (e.g., timber, fuel wood, fodder, fruit) and/or "services" (shelter, shade, soil or water conservation).

d) **Alley cropping:** It is a specific form of agroforestry in which trees and shrubs are grown in hedgerows within arable cropped land. Food crops are grown in alleys formed by hedgerows of trees or shrubs in arable lands. It is also known as "hedgerow" intercropping".

e) **Multi-storied cropping**: In this system perennial, seasonal, and or biennial crops of different height, structure and rooting pattern are grown in the same field at the same time. Examples are coconut, arecanut, black pepper, pine apple, banana, turmeric and ginger etc.

f) **Multi-layer cropping:** Annual crops of different heights and duration are frown in the same field at the same time. Examples are banana-okra-cowpea.

Advantages of Multiple Cropping

- Higher gross return per unit of land
- Reduces risk of natural calamities and pest attack etc.
- Stabilises yield and provides food security

- Inclusion of legumes enhances soil fertility
- Suppresses weeds and reduces soil erosion
- Better utilisation of resources and labour

Disadvantages of Multiple Cropping

- Difficulty in agricultural operations
- Allelopathic effect may reduce the yields
- There is no resting period for land
- Difficulty in farm mechanisation

7.3.3 Crop rotation

In a crop sequence, if different crops are grown in a sequence it is called crop rotation. The exact sequence of crops will vary depending on local circumstances, with the critical design element being an understanding what each crop contributes and takes from the soil. Crop rotation may be for one year, two years or three years. The crops to be grown in a rotation is carefully designed to ensure that soil nutrients are sustained, pest populations are controlled, weeds are suppressed and soil health is built. An ideal crop rotation makes the land both more productive and more environmentally sustainable. It improves the financial viability of a farm by increasing productivity whilst reducing chemical input costs.

Key advantages of crop rotation

Increase in soil fertility: Crop rotation improves the physical and chemical conditions of soil and thus improves the overall fertility. Legume crops fix atmospheric nitrogen and enrich soil fertility. The crops with deep rooted system can extract nutrients and water from deeper zones of soil making those available for shallow rooted crops grown subsequently.

Control of pests/diseases: Crop rotation helps to control common root and stem diseases that affect row crops. Crop rotation can be used to manage some non-mobile insects whose larvae or eggs overwinter in soil and which have a selected range of crops to feed on.

Increased organic matter content: Crop rotation will add more crop residues, green manures and other plant debris to the soil. Crop rotation also requires less intensive tillage, which means that soil organic matter does not degrade as quickly. Increased soil organic matter improves soil infiltration and water holding capacity, which enables water to be absorbed into the soil. Furthermore, increased of soil organic matter improves overall soil structure and the chemical and biological properties of the soil. Crop rotation helps control the erosion of soil from water and wind by improving the soil structure and reducing the amount of soil that is exposed to water and wind. Crop rotation also supports reduced or no-till farming, which ensures even better protection against erosion.

Reduction of soil erosion: Cover crops are effective in reducing raindrop impact, reducing sediment detachment and transport, slowing surface runoff, and so reducing soil loss. More flexible rotations should be designed for the regions that are susceptible to unseasonal rains or drought.

Soil Biodiversity: Crop rotation helps improve soil biodiversity by changing crop residue and rooting patterns. The microbial community in soil is supported by rotating crops with a high carbon to nitrogen ratio (such as corn) with low carbon to nitrogen ratio crops (such as soybeans).

Crop yield: Crop rotation can help increase yield of all crops included in the rotation.

7.4 Integrated Farming Systems (IFS)

Since income from cropping alone in small and marginal farms is insufficient to sustain the farmers' family there is a need to increase income from other enterprises like livestock, fishery, agro-forestry, apiary, sericulture, poultry, duckery etc. in an integrated farming system. Integrated farming is defined as biologically integrated farming system which integrates natural resources and regulation mechanism into farming activities to achieve maximum replacement of off-farm inputs; secure sustainable production of high quality food and other products through ecologically preferred technologies; sustain farm income; eliminates or reduces sources of present environmental pollutions generated by agriculture, and sustains the multiple functions of agriculture (IOBC, 1993).

The rationale behind integrated farming is to minimize wastes from the various subsystems on the farm. In IFS, the by-products from each subsystem are used as inputs to other subsystems to improve the productivity and lower the cost of production of the outputs of the various subsystems (Edwards et al. 1988, Gill et al. 2009). Integrated farming system increases the economic yield per unit area per unit time by virtue of intensification and diversification of crops and integration of allied enterprises. It also offers enough scope to nutrient recycling within the system to economize and sustain the system and minimizes the dependence on chemical fertilizers for crop production to earn more profit (Rangaswami et al. 1999, Ganesan et al. 1999). Overall an integrated farming system fulfill the multiple objective of making farmers self-sufficient by ensuring the family members a balance diet, improving the standard of living through maximizing the total net returns and provide more employment, minimizing the risk and uncertainties and keeping harmony with environment (Mali et al. 2014).

Goals of IFS

The goals of integrated farming systems (IFS) are to:

- To provide a steady and stable income and rejuvenation/ amelioration of the system's productivity.
- To achieve agro-ecological equilibrium through the reduction in the build-up of pests and diseases, through natural cropping system management and the reduction in the use of chemicals (in-organic fertilizers and pesticides).
- To provide environmentally sustainable and economically viable technology that encompasses rational utilization of available resources of the region.
- To conserve natural resource base, protect the environment and enhance prosperity for a longer period of time.

7.4.1 Integration of enterprises

Since IFS is an interrelated complex matrix of soil, water, plant, animal and environment and their interaction with each other it enables the system to be more viable and profitable over arable farming system and leads to production of the quality food. Judicious combination of enterprises, keeping in view of the environmental conditions of a locality will pay greater dividends. At the same time, it will also promote effective recycling of residues/wastes (Kumar et al. 2012a). Integration of enterprises includes Rice-fish integration, Horticulture-fish system, Mushroom-fish system and Sericulture-fish system. Livestock-fish system includes Cattle-fish system, Pig-fish system, Poultry-fish system, Duck-fish system, Goat-fish system, Rabbit-fish system etc.

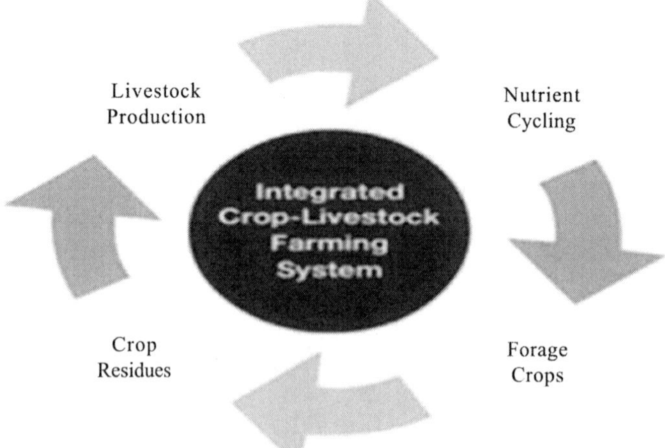

Fig.7.1. Integrated crop-livestock farming system

Enterprises of IFS

- **Crop husbandry**
- **Dairy**
- **Piggery**
- **Poultry**
- **Duck farming**
- **Aquaculture**
- **Fruit cultivation**
- **Vegetable production**
- **Agro-forestry**
- **Mushroom production**
- **Apiculture**
- **Sericulture**
- **Biogas plants**

Fig.7.2. Various enterprises in IFS

IFS can be practiced in different ways with variable intensity depending on socio-economic structure, characteristics of soil, choice of the farmers and most importantly the resource availability of the farmers. Generally it is adopted in three different farming situations, such as wetland, garden land and dry land farming situations. The combinations of enterprises will vary depending on the farming situation.

Wetland system	Rice-based system + fish culture+ duck/poultry
Garden land system	Garden crops with forage + livestock/poultry (crops with fodder, mulberry + dairy + silkworm rearing)
Dry land system	G + agroforestry (rain/fodder crops + sheep/goat rearing + agroforestry (tree crops/ agri-silvi/agri-horti.)

Different farming system models developed for varied agro-ecological regions are given in Table 7.1.

Table 7.1. Integrated farming system models for different agro-ecological regions of India

S.No.	Farming system model	Suggested by
1.	Composite fish culture	Jayaseelan et al. (1988)
2.	Duck cum fish farming	Sharma et al. (1985)
3.	Pig cum fish farming	Jhingaran and Sharma, (1980)
4.	Poultry cum fish farming	Sharma et al. (1985) Kumar and Singh (1984) Govindan (1988)
5.	Cropping cum mushroom	Rao (1984) Longathon (1989) Rangaswami and Jayanthi (1984)

6.	Rice + fish + poultry farming	Govindan et al. (1990)
7.	Rice + fish + poultry + mushroom	Rangaswamy (1992)
8.	Rice + fish +Dairy	Veerbudran (1994)
9.	Crop + Livestock	Sharma et al. (1991)

7.4.2 Principles of IFS

Integrated farming tries to imitate nature's principle, where not only crops but also varied types of plants, animals, birds, fish, and other aquatic flora and fauna are utilized for production. It enhances ecological diversity with minimum competition for water, nutrients and space between components.

7.4.3 Advantages of IFS

The approach aims at increasing income and employment from small-holding by integrating various farm enterprises and recycling crop residues and by products within the farm itself (Behera and Mahapatra, 1999, Singh et al., 2006). Diversification into farming system mode of agriculture on small land holding can provide proofing for predicted climate change related risk in agriculture. This can also help in obtaining food and nutritional security at farm level and can also generate rural employment, thus preventing excessive migration to urban areas, which is a common problem in developing economies (Singh, 2012). The other benefits are:

- It increases income and employment from small farms and produces diversified products.
- Improves soil fertility and reduces weeds, insect pests and diseases due to appropriate crop rotation.
- Utilizes of crop residues and livestock wastes.
- Reduces reliance on outside inputs – fertilizers, agrochemicals, feeds, energy etc.
- Increases net returns and provides diversified income sources, guaranteeing a buffer against trade, price and climate fluctuations (Kumar et al., 2015).

7.4.4 Scope for IFS

IFS gives greater importance for sound management of farm resources to enhance the farm productivity and reduce the environmental degradation, improve the living standard of resource poor farmers and maintain sustainability (Kumar et al. 2013). In IFS combination of one or more enterprises with cropping, when carefully chosen, planned and executed, gives greater dividends than a single enterprise, especially for small and marginal farmers. Farm as a unit is to be considered and planned for effective integration of the enterprises to be combined with crop production activity. Integration of farm enterprises to be combined on many factors such as: (i) soil and climatic features of the selected area, (ii) availability of resources, land, labour and capital, (iii) present level of utilization of resources, (iv) economics of proposed integrated farming system, and (v) managerial skill of the farmer.

7.4.5 Integrated farming system models

Crops, livestock, fish, birds and trees are the major components of any IFS. Crop may have subsystem like mono-crop, mixed/intercrop, multi-tier crops of cereals, legumes (pulses), oilseeds, forage etc. Livestock components may be milch cow, goat, sheep, poultry, bees etc. while tree components may include timer, fuel, fodder and fruit trees.

i) Rice–fish–livestock-horticulture-based farming system for rainfed lowland areas

NRRI, Cuttack has developed an adoptable technology of 'Rice-fish diversified farming system'. The farm size may vary from minimum of about one acre to one hectare or more.

Field design

Field design includes wide bunds (dykes) all around, a pond refuge connected with trenches on two sides (water harvesting come fish refuge system) and guarded outlet. The approximate area allotments will be, 20% for bunds, 13% for pond refuge and trenches and rest 67% for main field. The pond refuge measures 10 m wide and 1.75m deep constructed in the lower end of the field. The two side trenches of 3 m width and average 1 m depth have gentle(0.5%) bed slope towards the towards the pond refuge. Small low cost (Thatched/ asbestos top) duck house and poultry unit are constructed on bunds with a floor space of about 1.5 sq. ft. for each duck and 1 sq. ft. for each poultry bird. Poultry unit maybe projected up to 50% over the water in the pond refuge to utilize the dropping as fish food and manure in the system. In such case birds can be housed in cage of made of wire net.

Production technology

Production technology broadly involves growing of improved photo-period sensitive semi tall and tall wet season rice varieties tolerant to major insect pests and diseases. The suitable rice varieties are Gayatri, Sarala, CR Dhan 500, CR Dhan 505, Jalmani, Varshadhan, etc. for Odisha; Sabita, Jogen, Hanseswari,

Varshadhanetc. for West Bengal; Sudha for Bihar;Madhukar and Jalpriya for eastern Uttar Pradesh and Ranjit, Mashuri and Sabita for Assam. Management of insect-pests in rice crop is done with the use of sex pheromone traps, light traps and botanicals (Netherin/ Nimbicidin spray at 1%).

Fig. 7.3. NRRI model of IFS in lowland

Indian major carps i.e., Catla [*Catla catla*(Ham.)]; Rohu [*Labeo rohita*(Ham.)]; Mrigal [*Cirrhinus mrigala*(Ham.)]; exotic carps, common carp [*Cyprinus carpio*(L.)];silver carp [*Ctenopharyn godonidella*(Val.)];silver barb [*Puntius gonionotus*(Bleeker)] and fresh water giant prawn [*Macrobrachium rosenbergii*]fingerlings of 3-4" size and prawn juveniles of 2-3" size are released in a ratio of 75% and 25%, respectively at 10,000 per hectare of water area after sufficient water accumulation in the refuge and in the field. Fish and prawn are regularly fed at 2% of total biomass with mixture containing 95% of oil cake +rice bran (1:1) and 5% of fish meal. After rice, various crops like watermelon, green gram, sunflower, groundnut, sesame and vegetables are grown in the field with limited irrigations from the harvested rainwater. On bunds different seasonal vegetables are cultivated round the year including creepers on the raised platform, spices and pineapples are grown in shades. The fruit crops on bunds include varieties of dwarf papaya, banana T x D coconut and areca nut. Flowers like tuberose, marigold, etc. are also cultivated on the bunds. Both straw and oyster mushroom cultivation are done in the thatched or polythene enclose. Bee rearing is practice in 2-3 bee boxes on bunds. Agro-forestry component on the bund include short term plantation of mainly *Accacia mangium, Accacia auriculiformes*. Animal component constitutes improved breeds of duck, poultry birds and goats. Azolla is released at 0.5 -1.0 t/ha and is maintained to supplement duck feed and also to some extent fish feed, besides nutrition to the rice crop. Fresh water pearl culture is integrated in the system using the host mussel

(*Lamellidens marginalis*), which is normally available in the lowland rice ecology. Components can however, be included in the system based on location specific requirements.

Productivity and economics

The rice fish farming system can annually produce around 16 to 18 t of food crops, 0.6t of fish and prawn, 0.55 t of meat, 8000-12,000 eggs besides flowers, fuel wood and animal feed as rice straw and other crop residues from one hectare of farm. The net income in the system is around Rs. 76,000 in the first year. Subsequently, this increases to around Rs.1,30,000 in the sixth year. This system thus increases farm productivity by about fifteen times and net income by 20 folds over the traditional rice farming in rainfed lowlands. It also generates additional farm employment of around 250-300 man-days/hectare/year.

Table 7.2. Cost of production per ha

S.No.	Particulars	Cost (Rs)
1.	Construction of pond refuge and trenches and dykes (2000 m x 35)	70,000
2.	Constriction of platforms 16 No. @ 200/-	3200
3.	Pit digging, planting of fruit and silvicultural plants (125-130 No.)	4000
4.	Cost of seeds/seedlings/saplings	5000
5.	Cost of FYM/Vermi-compost	5000
6.	Cost of fingerlings	8000
7.	Cost of fish feed	5000
8.	Small farm implements/equipment	5000
9.	Labour 400 man days @ Rs. 150	60000
	Total	1,68,200

ii) Multi-tier rice-fish-horticulture-agro-forestry-based IFS for deep water

A multi-tier rice-fish–prawn horticulture crops-agro-forestry-based farming system model has also been developed in 0.06 hectares area at Institute Farm of NRRI, Cuttack for deep water (50-100 cm).

Design

The design of the system includes land shaping in the form of uplands (tier I and tier II) covering about 15% of field area followed by rice field area of 40% as rainfed lowland (tier III) and deep water (tier IV). This rice field is connected to a micro water shed-cum fish refuge (pond) of 20% area for growing fishes (catla, rohu, mrigal, silver carp, silver barb) and fresh water giant prawn along with the rice crop. Raised and wide bunds are made all around using 25% of the farm area.

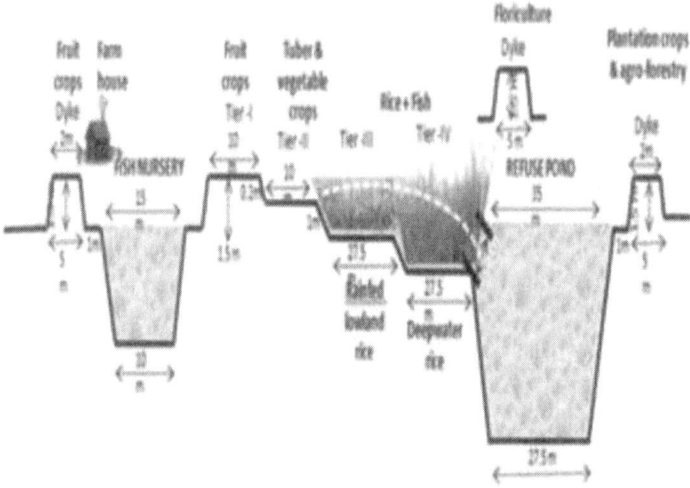

Fig.7.4 Field Design

Production technology

High yielding varieties of rainfed lowland rice (Gayatri, Sarala, Pooja) in tier III and deep water rice (Jayanti Dhan, CR Dhan 500, CR Dhan 505 CR Dhan 508, Pradhan Dhan and Varshadhan) are grown with the fish and prawn during wet season. Dry season crops like sweet potato, green gram, sunflower, groundnut, vegetables are grown after lowland rice in tier III. Dry season rice is cultivated after the deep water rice is harvested. Harvested rain water in the pond refuge is used for irrigation of the dry season crops. Improved varieties of perennial (mango, guava, sapota) and seasonal fruit crops (papaya, banana, pineapple, etc.) are grown in Tier I. Round the year different seasonal vegetables and tuber crops viz., sweet potato, elephant foot yam, yam bean, colocasia and greater yam are cultivated in Tier II. Agroforestry (Acacia mangium) and plantation crops (Coconut and areca nut) are planted on the northern side of the bunds. Greater yam is grown with the support of trunk of agro forestry tree. Poultry and duckery components are integrated on bunds of the pond refuge.

Productivity and economics

The total annually produce of about 14-15 t of food crops, 1 t of fish and prawn, 0.5-0.8 t of meat, 10000-12000 eggs in addition to flowers and 3-5 t of animal feed is harvested from 1 hectare farm area. The productivity of food crops further increases to 16-17 t besides, 10-12 t of fibre/fuel wood from eight year onwards due to addition of produce from perennial fruit crops and agroforestry components. The net income in this system is around Rs. 1, 00,000/ ha in the first year. This will increase to Rs 1, 50,000 or more from the eight year onwards.

iii) Rice-based farming system under irrigated lowlands

With the objective of improvement of livelihood of small and marginal farmers, rice-based integrated farming system model for irrigated areas has been developed at NRRI, Cuttack.

Production Methodology

About an acre of integrated farm area has been reoriented for the farming system of which 30% of the area is converted to two rice-fish fields of 600 sq. m area each with a refuge of 15% area and another 30% area is developed into two nursery fish ponds of equal size for fingerlings rearing. The remaining area is utilized as bunds for growing vegetables, horticultural crops and agro-forestry.

Fig.7.5. Rice based farming system

Three rice crops are grown in sequence of Kharif rice (var. CR Dhan 505/ Varshadhan) followed by Rabi rice (Naveen/High protein rice) and then summer rice (Vandana/Sidhant). Yellow stem borer is controlled by using pheromone traps or by applying 1% Nethrin/Nimbecidine. Fish culture is taken up with catla, rohu and mrigal species. The fish fingerlings are reared in the two nursery ponds and are used for culture with rice crop in the system. The excess fingerlings are sold out. On the bunds agro-forestry plants like teak, Accacia, Sisoo, Neem, Aonla and bamboo are planted on the northern and southern bunds. Horticultural crops such as banana, papaya and arecanut are grown on the bunds. Pineapple and spices are cultivated in the shade. Flowers like Marigold, Hibiscus and

Jasmine are also 2 cultivated in the western bund in 50m area. Two fruit plants of lemon ,guava, jackfruit, mango and litchi each are also planted on the southern bund near the farm house to meet the household requirement. One poultry and one duckery unit are integrated in the system in which 40 poultry birds are raised during the dry seasons (October to April) and 20 ducks are reared during the wet season (July to December).

Productivity and economics

Three crops of rice yield 800 to 1000 kg of grain per year. Entire produce is sufficient to cater the need of the small farm family. Paddy straw is used for production of mushroom, use as cattle feed and rest is sold. After two to three months of rearing, fish fry worth of Rs. 4000- 5000 is sold to other farmers. Fish is harvested according to the need after the size becomes 250-300 g after 6 months or 0.5-1.0 kg after a year. The income from fish rearing in the system is Rs. 20,000. Pulses (green gram, black gram and pigeon pea) taken on the slope and bunds are just enough to meet the protein requirement of the farm family.

iv) Integrated fish farming

Integrated farming approach can be adopted to boost food production by integrating aquaculture with dairy, ducks, poultry, pigs and horticulture. The systems recycle organic waste and economises the production cost. Pond bottom humus is used to fertilise crops grown adjacent to ponds or directly on their dykes. Livestock manures are used as pond fertilisers for production of plankton. The common integrated fish farming types are (i) paddy-cum fish culture, (ii) fish cum duck farming, (iii)paddy-fish-azolla farming, (iv) crop husbandry and fish culture, (v) fish cum poultry farming, (vi) fish cum pig farming, (vii) paddy-fish- vegetable farming, (viii) fish cum sericulture, (ix) fish cum horticulture and (x) fish cum biogas plant.

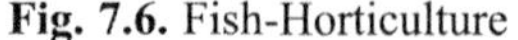

Fig. 7.6. Fish-Horticulture

Fig. 7.7. Fish –Duck farming system

7.5 Diversified farming system for sustainability

A diversified farm is one that has several production enterprises or sources of income. Diversified or bio-diverse ecosystems are more structurally complex and stable than the simplified monoculture-based systems. They are resistant to external disturbances including species invasions, diseases, and other disturbances.

Fig. 7.8. Diversified farming

More diverse ecosystems contain many species with similar function which is crucial in providing the stability in the ecosystems. Furthermore, highly bio-diverse communities may be able to tap resources more effectively because different species differ in strategies for resource acquisition. Studies have shown that beneficial soil microbial communities that support soil health are also greater in diversified organic farms, compared to monoculture-based conventional farms.

Agro-forestry

Agroforestry is a sustainable land-use system and technology whereby woody perennials are deliberately used on the same land management unit as agricultural crops and/or animals in some form of spatial arrangement or temporal sequence. An Agro-forestry system is more acceptable than tree farming alone, since the intercropped annuals give income when the tress are too young to yield beneficial produce. There are many types of agro-forestry systems depending on the type of crops grown.

Agri-silvicultural systems : Annual agricultural crops are grown between tree rows in such system as inter crops for two years. However fodder crops, shade loving crops and shallow rooted crops can be grown economically.

Fig. 7.9. Agro-silvocultural system

Silvi-pastoral system : In this system the trees and shrubs are grown with pasture for livestock. The trees and shrubs may be used primarily to produce fodder for livestock or they may be grown for timber, fuel wood and fruit or to improve the soil. This system is classified in to three categories such as (i) protein bank, (ii) live-fence of fodder trees and hedges, and (iii) trees and shrubs on pasture.

Fig.7.10. Silvo-pastoral system

Agri-silvipastoral system: The system integrates crop and /or animal with trees. Woody perennials having fodder value are introduced in the system. Such systems can be used for soil conservation besides producing food, fodder and fuel. It may be tree-livestock crop mix around homestead, wood-hedge rows for browsing, green-leaf manure and soil conservation or for an integrated production of pasture, crops animals and wood.

Agri-horticultural system: In this system short duration arable crops are raised in the interspaces of fruit trees. Some of the fruit trees that can be considered are guava, pomegranate, custard apple, sapota and mango. Pulses are the important arable crops for this system.

Tree-based farming system (WADI): This system is developed as a one acre mini-orchard with around 100 fruit trees of guava, mango/gooseberry (amla) or other varieties appropriate to the region. The space between the fruit trees is used for growing seasonal crops and the periphery is bio- fenced with forestry, fuel or timber species. From an individual farm perspective, it is a tree-based farming system, more specifically a Wadi system, in which the physical unit interacts with other production components of the farm such as annual crop fields and livestock. At the level of the physical land unit, the Wadi plot is an agri-horti-forestry arrangement of beneficial plant species.

In WADI system, the interaction among the components is in the tree-crop interface. For example, the fodder from the forestry species in the Wadi is used as fodder for livestock and the dung or farmyard manure is returned to the interspaces where annual crops are grown. Similarly, there is interaction among the Wadi and non-Wadi land in sharing labour and inputs. Incorporation of these factors in the design and execution is a major reason for the success of the Wadi programme.

Design

The WADI plot has agricultural, horticultural and forestry species in 0.4-1.0 ha of land. The arrangement of these species generally centres on the horticultural component where trees like mango, cashew and amla are planted at their recommended spacing, which ranges from 10 x 10 m to 6.0 x 6.0 m. As intercrops are grown in the interspaces of trees, their yield is additional and not at the expense of fruit / nut yield of the horticultural crops.

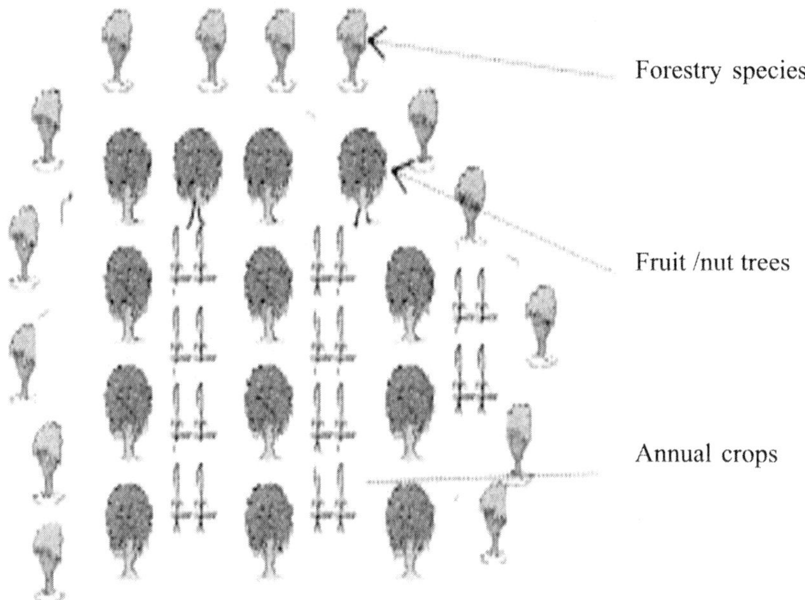

The multi-purpose forestry species like Subabul and Gliricidia. are planted at relatively close spacing along the border of the designated WADI plot. The shift from the rainfall-dependent single crop to at least three species in the WADI enhances the ecological sustainability of the farm. At the same time, the product diversity in the form of food, fodder, fuel wood and small timber increases the economic sustainability of the farmer.

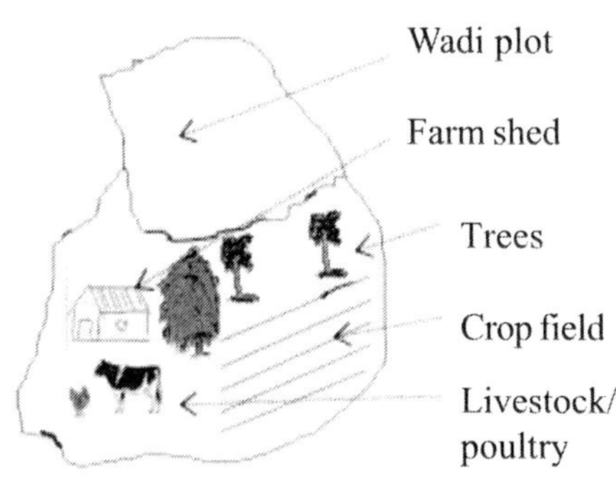

Fig. 7.11. Tree based WADI system

Sloping Area Land Technology (SALT)

Shifting cultivation of Jhooming is an old age practice in hilly areas. This is also known as "slash and burn" (swidden)) system of farming. Previously jhooming cycle was 10-20 years for regeneration of forest which has come down to 2-3 years as a result there is rapid depletion of its forest cover. Due to shifting in short cycles the forest areas of generally fragile, sloping soils, are subject to intensive agriculture practices, which rapidly degrade the land. There is severe soil erosion along with deforestation of forest due to such type of cultivation.

There are several traditional ways of controlling soil erosion such as reforestation, terracing, multiple cropping, contouring and cover cropping. The Asian Rural Life Development Foundation (ARLDF) promotes an erosion control technique that is both easier and less expensive to implement than the traditional methods. This technology is known as SALT or "Sloping Agricultural Land Technology".

SALT is a package technology of integrated soil conservation and food production. Field and permanent crops are grown in 3-5 m wide strips between contoured rows of nitrogen fixing trees. For the purpose of creating thick hedge rows for soil conservation, the nitrogen fixing trees are thickly planted in double rows. When the hedge plants grow 1.5-2 m tall, they are cut down to about 75 cm and the cuttings (tops) are placed in the alleyways to serve as green manure.

SALT is an agroforestry since rows of permanent shrubs like coffee, cacao, citrus and other fruit trees are dispersed throughout the farm plot. The strips not planted by permanent crops, are planted alternately with annual crops like cereals (corn, upland rice, sorghum, etc.) or other crops (sweet potato, melon, pineapple, castor bean, etc.) and legumes (soybean, mung bean, peanut, etc.). This cyclical cropping provides the farmer some harvest throughout the year. Planting of trees for timber and firewood (Casurina, Sesbania, Cashew) can be taken on surrounding boundaries.

Advantages of SALT

- It is a simple and low-cost and timely method of upland farming.
- It uses low cost technology as contour lines are determined by using an A-frame transit that any farmer can learn to make and use.
- Varieties of crops with and old farming patterns can be utilized in the SALT system.
- The nitrogen-fixing trees can enrich the soil fertility.
- It is a resource conservation method of cultivation which reduces soil erosion.

Various Forms of SALT

There are four types of SALT which are numbered as SALT-1, SALT-2, SALT-3 and SALT-4. SALT-1 is the most important form of diversified farming which follows the principles of agroforestry.

SALT 2 (Simple Agro-Livestock Land Technology) is a small, livestock-based agroforestry system (preferably with dairy goats) and has a land use of 40% for agriculture, 20% for forestry and 40% for livestock. The hedgerows of different nitrogen fixing trees and shrubs are established on the contour lines and they use manure from the animals is utilized as fertilizer both for food and forage crops.

SALT 3 (Sustainable Agro-forest Land Technology) is a cropping system in which a farmer can incorporate food production, fruit production and forest trees that can be marketed. The farmer first develops a conventional SALT project to produce food for his family and possibly for livestock and then he can plant fruit trees between the contour lines. The plants in the hedgerows are cut and piled around the fruit trees for fertilizer and soil conservation purposes. A small forest of about 1 ha will be developed in which trees of different species are grown for short-range production of firewood and charcoal.

SALT 4 (Small Agro-fruit Livelihood Technology) is based on a half-hectare piece of sloping land with two-thirds of it developed in fruit trees and one-third intended for food crops. Hedgerows of different nitrogen-fixing trees and shrubs are planted along the contours of the farm.

The ten steps of SALT

SALT adopts 10 simple steps in its life cycle.

Step 1: Make an A-frame

The first step is to make an A-frame which is a simple tool made out of locally available materials. Three sturdy wooden or bamboo poles, a saw or bolo, an ordinary carpenter's level, and string or rope are used for preparing the A-frame as shown in the figure. The A-frame is used to find the contour lines of the land. The contour line is a level line from one end of the field to the other and is found around the hill or mountain.

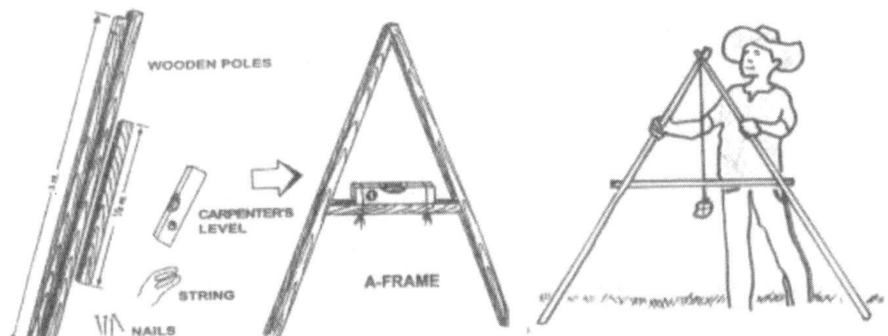

Fig. 7.12. Preparation of A-frame

Step 2: Locate and mark the contour lines

After locating and marking the contour lines at certain intervals tall grasses or any obstruction are to be removed. Make a study of the area for which contour lines are to be determined. Begin marking contour lines near the highest point. Let the A-frame stand on the ground. Without moving the rear leg, then put the front leg down on the ground that is on the same level with the rear leg.

There are two criteria for determining the distance between contour lines: vertical drop and surface distance. Generally, no more than a 1 m vertical drop is desirable for effective erosion control. In steeper slopes the contour lines are closer and wider in flatter slopes.

Fig. 7.13. Laying out a contour line

Step 3: Prepare the contour lines

After the contour lines are located , ploughing and harrowing are done along the contour lines for planting. The width of each area to be prepared should be 1 m.

Fig. 7.14. Ploughing along contour lines

Step 4: Plant seeds of nitrogen-fixing trees and shrubs (NFTS)

Two furrows at a distance of 0.5 m apart are made on each prepared contour line for sowing seeds. Seeds are sown and covered lightly and firmly with soil. The ability of NFTS to grow on poor soils and in areas with long dry seasons makes them good plants for restoring forest cover to watersheds, slopes and other lands that have been denuded of trees. Through natural leaf drop they enrich and fertilize the soil. Subabul and Gliricidia are preferred NFTS.

Step 5: Cultivate alternate strips

Alternate strips are cultivated to prevent soil erosion as the unploughed strips will hold the soil in place. When the NFTS are fully grown, every strip can be cultivated.

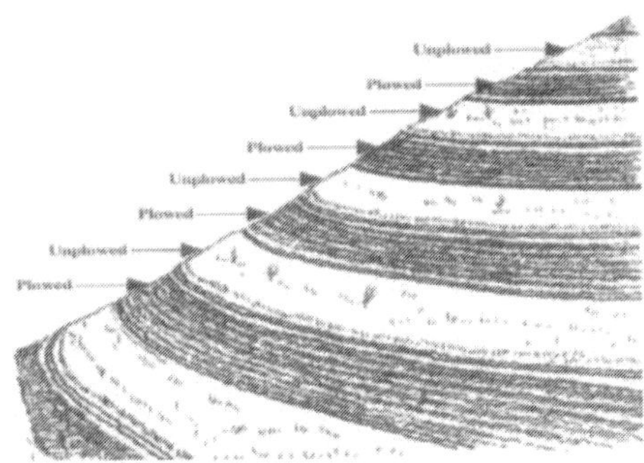

Fig. 7.15. Cultivating alternate strips

Step 6: Plant permanent crops

Every third strip can be planted with permanent crops at the same time the seeds of NFTS are sown. Only the planting holes are cleared and dug and ring weeding is done afterwards. Permanent crops like coffee, banana, citrus, cacao, and others of the same height are usually preferred.. Tall crops are planted at the bottom of the hill while the short ones are planted at the top. Shade-tolerant permanent crops can be intercropped with the tall crops.

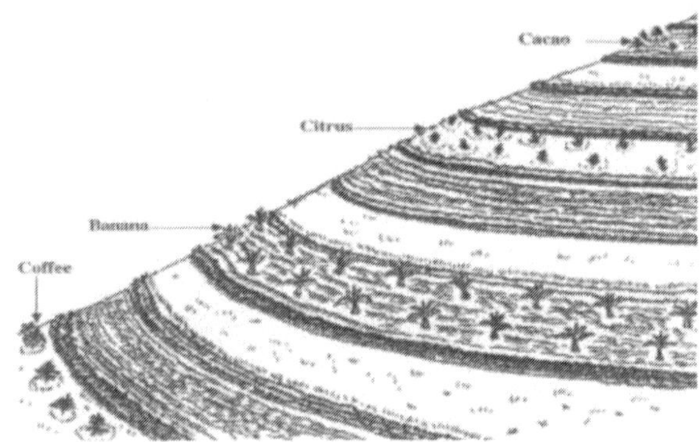

Fig. 7.16. Planting permanent crops

Step 7: Plant short- and medium-term crops

Short and medium duration crops can be planted between and among strips of permanent crops to serve as a source of food and regular income till the permanent crops bear fruits. Suggested short and medium-term crops are pineapple, ginger, castor bean, peanut, mung bean, melon, sorghum, corn, upland rice, etc.

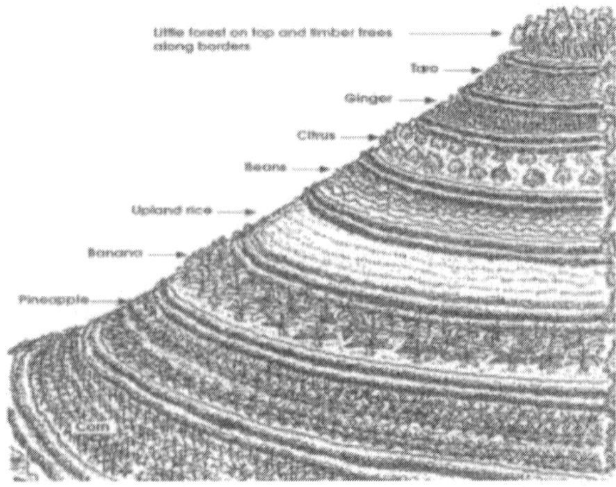

Fig. 7.17. Alternate strips of short-term and medium-term crops

Step 8: Regularly trim the NFTS

The NFTS are cut back to a height of 0.5-1 m from the ground once in a month. Pruned leaves and twigs should always be piled at the base of the crops. They serve as soil cover to minimize the impact of raindrops on the bare soil. They also act as excellent organic fertilizer for both the permanent and short-term crops.

Step 9: Practice crop rotation

Crop rotation of annual field crops should be practiced involving cereals, legumes, vegetables and oilseeds. Other management practices in crop growing, like weeding and pest control, should be done regularly.

Step 10: Build and maintain green terraces

Control of soil erosion is done by the double-thick rows of nitrogen-fixing trees and the natural terraces being formed along the contour lines of the hill. Mulching is also done for soil and moisture conservation by gathering and piling up straw, stalks, twigs, branches, leaves, rocks, and stones at the base of the rows of nitrogen fixing trees.

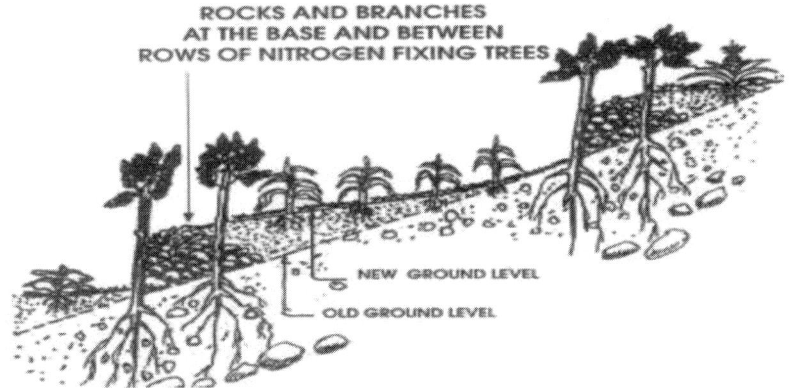

Fig. 7 18. Building terrace

7.6 Sustainability through IFS

Sustainable development in agriculture must include integrated farming system (IFS) with efficient soil, water crop and pest management practices, which are environmentally friendly, and cost effective . Nutrient recycling within the system advocates the self-sustainability of the system and which will not only reduce the dependency on the external inputs viz., seed/ fertilizers etc. but also provide the balanced and rich nutrition to the farm family with reduced cost of cultivation and increased profit margin on the same piece of land which is key factor for taking care of sustainability. On any farm, four natural ecosystem processes like energy flow, water cycle, mineral cycle and ecosystem dynamics work (Sullivan 2003). These four ecosystems processes function together, complementing each other as sustainable agriculture requires system approach (Singh et al. 2009) and system implies a set of agricultural activities organized while preserving land productivity and environmental quality and maintaining a desired level of biological diversity and ecological stability.

References

Behera U K and Mahapatra I C. 1999. Income and employment generation of small and marginal farmers through integrated farming systems. Indian Journal of Agronomy 44(3): 431–9

Dipak De and Basavaprabhu Jirli. 2017. Entrepreneurship in agricultural development. Daya pub., New Delhi,India, pp.126

Edwards P, Pullin R S V and Gartner J A. 1988. Research and KUMAR ET AL. November 2018] 1673 25 and employment generation options for small and marginal farmers. PDCSR, Modipuram, pp 109-18

Ganesan G, Chinnasamy K N, Bala Subramanian A and Manickasundram P. 1990. Studies on rice-based farming system with duck cum fish culture in deltaic region of Thanjavur district, Tamil Nadu. Farming Systems Newsletter 1(2) : 14

Gill M S, Singh J P and Gangwar K S. 2009. Integrated farming System and agriculture sustainability. Indian Journal of Agronomy 54(2): 128–39

Himani, B. Patel. 2016. Precision farming – a new technique for horticultural crops. Innovative Farming, 1(2): 35-37.

IOBC/WPRS, 1993: (A. El Titi, E.F. Boller & J.P. Gendrier, Eds.): Integrated Production: Principles and Technical Guidelines (in English, French, German). – IOBC/wprs Bull. 16 (1)

Kumar Sanjeev, Singh S S, Kumar Ujjwal, Dey A and Shivani. 2013. Sustainability of integrated farming systems in the eastern region. Natural Resource Conservation-Emerging Issues and Future Challenges, pp. 455-64. Madhu M, Jakhar Praveen, Adhikarry P P, Gowda Hombe, Sharda V N, Mishra P K and Khan M K (Eds). Kumar Sanjeev, Bhatt B P, Dwivedi S K and Shivani. 2015. Family farming- A mechanism to contribute towards global food security. Indian Farming 64(12): 13-8.

Kumar Sanjeev, Singh S S, Meena M K, Shivani and Dey A. 2012a. Resource recycling and their management under integrated farming system for lowlands of Bihar. Indian Journal of Agricultural Sciences 82(6): 504-10

Mali Hansram, Kumar Amit and Katara Pawan. 2014. Integrated farming system for irrigated and rainfed conditions. (In) Proceedings of National Symposium on

Agricultural Diversification for Sustainable Livelihood and Environmental Security, held during 18-20 November 2014 at Ludhiana, Punjab, p 546

Mindanao Baptist Rural Life Center Editorial Staff (2012): Sloping Agricultural Land Technology

Rangaswami A, Manicksundaram P and Vidhya G. 1999. Integrated farming system: a variable approach. Farming Systems News Letter 1(2): 11–3.

Rathore, E and Panda, S (2018): Diversification of Agriculture: A Framework of Agriculture Business for Linking Farming to Market.

Singh Gurbachan. 2012. Integrated farming systems: option for diversification to manage climate change related risk and livelihood security. (In) lead papers Vol -I Third International Agronomy Congress, held during 26-30 November at New Delhi, pp 93-4.

Singh S P, Gangwar B and Singh M P. 2009. Economics of farming systems in Uttar Pradesh. Agricultural Economics Research Review 22 (January-June): 129-38.

Singh V P and Rai S C. 2006. Integrated Farming System. World Food Day: 31-6

Sullivan P. 2003. Applying the Principles of Sustainable Farming: Fundamentals of Sustainable Agriculture. ATTRNA- National Sustainable Agricultural Information Service, Fayetteville, USA.

8

Sustainable Natural Resource Management

8.1 Natural resources

Natural resources are naturally occurring resources which can be exploited for economic gain. They include raw materials such as fuels, minerals and metals, soil, water, air, sunlight, land, biomass and ecosystems. Natural resources also include living (biotic) and non-living (abiotic) materials. Natural resources are gifts of nature for the humankind. These resources are required for existence of all animals and human beings. They may be grouped as (i) renewable and (ii) non-renewable. Air, water, sunlight, forests etc. are renewable as they can be replenished in a short period. On the other hand, coal, fossil fuel, natural gas, mineral etc. which cannot be replenished in a shorter period and proved to be finite resources are called non-renewable resources.

Natural resources like soil, forest, water and air provide fundamental support to life and economic processes. Soils are the basic building block in the livelihoods of people while forests protect water sources, biodiversity, reduce the risks of natural disasters and act as major carbon sink that mitigates climate change. Ecological processes maintain soil productivity, recycle nutrients, biological cycles, cleanse air and water, and regulate climatic cycles. Natural resources provide the foundation for maintaining and improving the quality of life of the population and can make greater contributions to sustainable growth. But these require proper management and any mismanagement coupled with a growing population, higher levels of economic activity per capita, and the complex interactions of these phenomena can put pressure on this foundation.

Soil degradation like erosion, salinization, compaction, and other forms of degradation affect 30 per cent of the world's irrigated lands, 40 per cent of rainfed agricultural lands, and 70 per cent of rangelands. The world's oceans are threatened by nutrient, plastic and heavy metal pollution, severe overfishing, and disease. Degradation of the natural resource base is having a substantial impact on the economies of developing countries. Degradation of the natural resource base also threatens long-term growth of a country mainly due to loss of agricultural production. Improving agricultural productivity is an essential part of development and poverty alleviation strategies in many countries and these objectives are often threatened by degradation of soil and water resources.

Fig. 8.1. Natural resources

Agriculture and natural resources are viewed to be not only the context of food production, but they are the main resources of small-scale rural livelihoods. National resources are viewed as natural capitals of rural households and communities' livelihoods in the framework of Sustainable Rural Livelihood (Fabricius, Koch, Magome, & Rurner, 2004).

8.2 Natural resources of India

Land resources: The geographical area of 328.73 M ha is broadly grouped into three sections (i) agricultural sector (59.27%) consisting of net cultivated areas, current follows, other follows and cultivable wastes; (ii) ecological sector (33.56%) comprising forests, miscellaneous, barren and uncultivable waste, and (iii) non-agricultural sector (7.17%) includes land under non-agricultural uses. The predominant soil groups of India are: red (105.5 million ha), black (73.5 million ha), alluvial (58.4 million ha), laterite (11.7 million ha), desert (30 million ha) and hill and terai soils (26.8 million ha). The per capita availability of land for cultivation is constantly declining (estimated to decrease to 0.17 ha in 2000 to 0.14 ha by 2025) and there is very little scope for horizontal expansion of net cultivated area.

Water resources: Rain is the ultimate source of water which is about 400 m ha m per year. The details of water resources in India are given in Table 8.1.

Table 8.1. Water resources in India

Particulars	Units
Annual rain water received in India	400 M ha m
Irrigation potential of India	139.5 m ha
Major and medium schemes	58.5 m ha
Minor irrigation schemes	15.0 m ha
Ground water exploitation	66.0 m ha
Utilization of irrigation potential	78.5 m ha
Net irrigated area	57.24 m ha

India has created an irrigation potential of about 84.9 M ha against the ultimate irrigation potential of 113.5 M ha (now revised to 139.5 M ha). It estimated that even after achieving the full irrigation potential, nearly 50% of the total cultivated area would remain rainfed.

Vegetation resources: Currently an area of 70 million ha is covered by forest in India. Per capita forest availability is only 0.07 ha. Deforestation has been one of the major causes of land degradation, with far reaching consequences to humanity. The poor quality of forest cover has not only accelerated the problems of environmental degradation but has also led to the deficit of fuel and fodder, 76 million tonnes dry fodder (Anonymous, 2000).

8.3 Natural resource management

Natural resource management (NRM) refers to the sustainable utilization of natural resources such as land, water, air, minerals, forests, fisheries, and wild flora and fauna to enhance human welfare. Sustainable NRM and poverty alleviation are usually highly compatible. The poor are heavily dependent on natural resources for their livelihoods while they are vulnerable to the effects of natural resource degradation. Hence natural resource management is essential for welfare of the poor and vulnerable communities.

Natural resource development refers to the development of natural resources such as land, water, soil, plants and animals with a particular focus on how management affects the quality of life for both present and future generation.

Over the past few decades, biological resources are being so intensely extracted and used that they do not have time to regenerate adequately. Anthropogenic pressure on these resources has reached critical levels such that not only productivity but also the above-ground biomass needed to maintain the resource is rapidly consumed. The same holds true for water which is an intrinsically recyclable resource, even soils which are naturally renewable due to their natural dynamic process. In view of the population growth, water availability has become insufficient in many countries worldwide.

8.4 Key issues of natural resource management

Land: Agriculture land is under tremendous pressure as it is limited and shrinking which is at constant risk of being further degraded. Land degradation affects agricultural productivity which becomes a major factor for food security and elimination of rural poverty. Doubling food production by 2050 to meet human needs will cause tremendous pressure on such degraded lands.

Sustainable land resources management (SLRM) is needed to protect and enhance the productive base of land resources and the livelihoods of the people, who are dependent on them. An efficient land policy framework is needed, including security of land rights and land access, establishment of the institutional infrastructure to administer land rights, and facilitation of land markets and transferability of land rights. SLRM hinges on a new approach of agricultural intensification that combines three basic principles: integrating the biophysical and socioeconomic driving forces involved; fostering a people-centred learning and participatory approach; and bringing recognizable and early productivity benefits to farmers.

Forests: Forests play an important role in poverty alleviation, sustainable economic growth, and the provision of ecosystem services. Indiscriminate deforestation causes natural imbalance in ecosystem. The three basic objectives of the new forest strategy are closely linked with the key objectives of the (i) harnessing forests to reduce poverty, (ii) integrating forests into sustainable economic development, and (iii) protecting global forest values. The three goals are to:

1. Harness the potential of forests to reduce poverty by creating opportunity, empowerment, and security for rural people, especially the rural poor and indigenous groups, in the use and management of forests.
2. Integrate forests into sustainable economic development. The approach is based on the fact that forests are seriously undervalued and are utilized wastefully and unsustainably in many economies, largely as a result of governance failures and perverse incentives.
3. Protect vital global forest values. The most important challenge in this area is to create effective markets for global values and other externalities from forests so that local and national stakeholders will benefit from protecting and managing the resource.

Water: Unsustainable use and management of water resources have caused a systemic water crisis in the world. The world's major lakes, rivers, and aquifers are under severe stress. Over population, urbanisations, industrialisation, over extraction of ground water and pollution have posed new threats to water sector.

The key future challenges include promoting a sound institutional environment; improving economic analysis of management options; improving transboundary water management; addressing social and sustainability issues in new dam construction; halting degradation and loss of ecosystem functions and the deterioration of freshwater lakes and reservoirs, wetlands, mangroves, and coral reefs; improving drainage; and addressing the water resources implications of climate change.

Biodiversity : The vast array of the world's animals and plants, the genetic information they contain, and the dynamic and interacting communities they form are known collectively as biodiversity. Biodiversity therefore permeates all levels of NRM, since its individual elements interact in intricate ways to form forests and grasslands, maintain soils, and provide ecosystem services, among other fundamental functions. Biodiversity provides two special challenges for NRM: (a) most of its benefits are economic externalities, that is, they do not appear as financial values on a market where they can be easily observed, and (b) some benefits of biodiversity accrue over the long term, while the cost of conservation may be more immediate. Another consideration is that many people consider biodiversity as having intrinsic value, for moral, religious, or cultural reasons. The perception that biodiversity is a global issue stems from the fact that its widespread decline has cumulative consequences at the global level.

8.4 Sustainable land management

Land is basic requirement for agricultural production. It also is an essential condition for improved environmental management, including source/sink functions for greenhouse gasses, recycling of nutrients, amelioration and filtering of pollutants, and transmission and purification of water as part of the hydrologic cycle. Therefore sustainable land management is also important for sustainable agriculture.

Sustainable land management is the use of land to meet changing human needs (agriculture, forestry, conservation), while ensuring long-term socio-economic and ecological functions of the land. The objective of sustainable land management (SLM) is to harmonise the complimentary goals of providing environmental, economic, and social *opportunities* for the benefit of present and future generations, while maintaining and enhancing the *quality* of the land (soil, water and air) resource (Smyth and Dumanski, 1993).

8.5 Soil and water conservation

Soil and water are basic resources to the requirement for food, feed, drinking, fuel, and fibre of human beings. Worldwide, around 52% of total productive land has been degraded by various kinds of degradation processes and almost 80% of the terrestrial land is affected by water erosion. Further, annually nearly 10 million hectares of cropland becomes unproductive at the global level due to soil erosion with an average rate of 30 t ha^{-1} $year^{-1}$. On the other hand, the future of living beings and agricultural production systems is at stake due to continuously depleting aquifers and increasing pressure on underground water under projected climate change scenarios. Moreover, climate change will increase water demand globally by about 40% of the water needed for irrigation .Hence, under the emerging scenario of acute water shortages and land degradation there is need for development and adoption of efficient approaches for soil and water conservation for agricultural sustainability.

8.5.1 Soil erosion

Soil erosion is a process of detachment and displacement of soil particles from land surface. Soil erosion is caused by a natural phenomenon, called geological erosion, and other human activities that lead to air and water erosion. In accelerated erosion, the rate of soil erosion exceeds a certain threshold level and becomes rapid. The human activities responsible for soil erosion are slash and burn agriculture (shifting cultivation), overgrazing of animals, deforestation, mining and faulty agricultural practices.

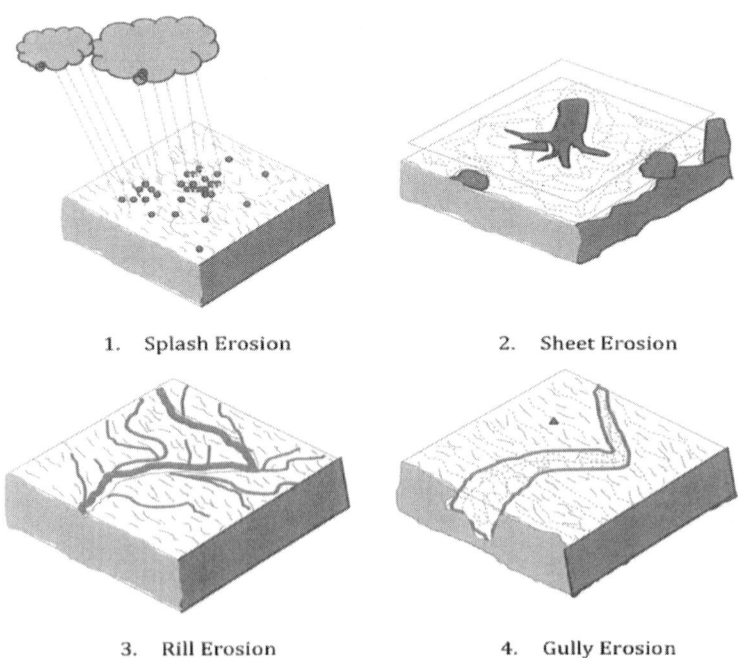

Fig. 8.2. Types of soil erosion

Globally, around 1100 m ha is affected by water erosion (56% of the total degraded land) and around 28% of the total degraded land area is affected by wind erosion. Of the geographical area of 329 m ha in India, 68 m ha is estimated critically degraded and another 107 m ha is severely eroded.

Rainfall is one of the major factors causing soil erosion. Excess rainfall that cannot infiltrate into soil runs off. Splash, sheet, rill and gully erosion are main forms of soil erosion by water (Fig. 8.2). The other forms of water erosion are ravine formation, slip, tunnel, stream bank, and coastal erosion.

Impact of soil erosion on agriculture

Higher soil erosion results in the removal of fertile topsoil along with nutrients which leads to reduced agronomic yield, land degradation, and terrain deformation. According to an estimate of existing soil loss data, the mean annual rate of soil erosion in our country is approximately 16.4 ton ha^{-1}. Soil erosion also removes soil organic matter (SOM) which is vital for improving soil bio-physico-chemical properties of soil. The soil removed by erosion has 1.5–5 times higher SOM than the soil left behind. The availability of SOM also affects the biological activities and soil biodiversity in a particular agro-ecosystem. The comprehensive impacts of erosion on soil and water resources are liable to reduce agricultural productivity (Fig. 8.3).

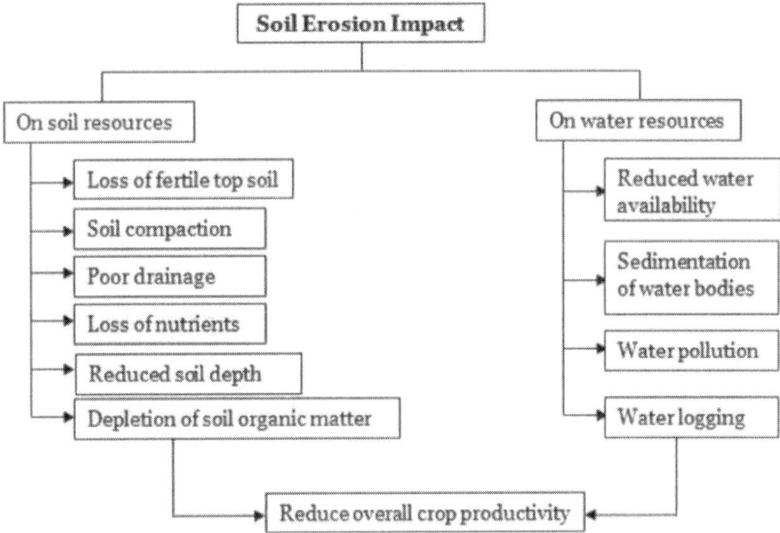

Fig. 8.3. Impact of soil erosion

To sustain agricultural productivity, it is imperative to reduce runoff, soil loss, and nutrient loss through water erosion. Other processes of land degradation are soil compaction, waterlogging, acidification, alkalinisation, and salinization depend on parent material, climatic conditions, and crop management practices.

8.5.2 Soil and water conservation measures

The measures required for soil conservation are combination of controlling or preventing soil erosion and maintaining its fertility. The conservation efforts should be to control erosion sufficiently, maintain soil organic matter content, maintain soil physical properties and maintain appropriate levels of nutrients in the soil. There are three types of measures for soil and water conservation, that is, mechanical, biological and agronomical.

Agronomical or biological measures are cheaper, sustainable, and may be more effective than structural measures, sometimes .Important agronomic measures are contour farming, choice of crops for appropriate soil and water conservation, crop rotation, cover cropping, inter cropping, strip cropping, mulching, conservation tillage, organic farming, land configuration techniques, and agro-forestry practices.

8.5.3 Soil health management

Soil health is defined as the continued capacity of soil to function as a vital living ecosystem that sustains plants, animals, and humans. Healthy soil gives us clean air and water, bountiful crops and forests, productive grazing lands, diverse wildlife, and beautiful landscapes. Soil does all this by performing five essential functions:

- *Regulating water* – Soil helps control where rain, snowmelt, and irrigation water goes. Water flows over the land or into and through the soil.
- *Sustaining plant and animal life* – The diversity and productivity of living things depend on soil.
- *Filtering and buffering potential pollutants* – The minerals and microbes in soil are responsible for filtering, buffering, degrading, immobilizing, and detoxifying organic and inorganic materials, including industrial and municipal by-products and atmospheric deposits.
- *Cycling nutrients* – Carbon, nitrogen, phosphorus, and many other nutrients are stored, transformed, and cycled in the soil.
- *Providing physical stability and support* – Soil structure provides a medium for plant roots. Soils also provide support for human structures and protection for archaeological treasures.

Soil management encompasses a number of strategies used by farmers to protect soil resources, one of their most valuable assets. These soil conservation methods allow for healthy soil formation, soil fertility and favourable soil composition, including soil permeability and soil porosity, which lead to increased soil health. Application of soil organic matter, cover cropping and crop rotation can be adopted for improving soil health.

8.6 Biodiversity conservation

The term biodiversity refers to the variety of life on Earth at all its levels, from genes to ecosystems, and can encompass the evolutionary, ecological, and cultural processes that sustain life. There are three types of bio-diversity such as (i) genetic diversity, (ii) species diversity, and (iii) ecological diversity.

Biodiversity is important to most aspects of our lives. We get many basic needs from biodiversity such as food, fuel, shelter, and medicine. Further, ecosystems provide crucial services such as pollination, seed dispersal, climate regulation, water purification, nutrient cycling, and control of agricultural pests. Biodiversity also holds value for potential benefits not yet recognized, such as new medicines and other possible unknown services. Biodiversity has cultural value to humans as well, for spiritual or religious reasons for instance. The intrinsic value of biodiversity refers to its inherent worth, which is independent of its value to anyone or anything else. This is more of a philosophical concept, which can be thought of as the inalienable right to exist. Finally, the value of biodiversity can also be understood through the lens of the relationships we form and strive for with each other and the rest of nature.

The total number of living species identified in India so far is 1,50,000 out of total twelve bio-diversity hot spots in the world. India has 15% of world's livestock with 2.4% of geographical area, 1% of forest area, and 8.5% of pasture land. Over 45,000 species of plants are found in India (17,000 flowering and 28,000 non-flowerings).

Loss of bio-diversity is a major concern now. This is caused due to loss of their habitats or destruction of some species. Natural causes include floods, earthquakes, landslides, natural competition between species, lack of pollination and diseases.

Destruction of habitat in the wake of developmental activities like housing, agriculture, construction of dams, reservoirs, roads, railway tracks, etc. are called man-made causes. Some of the important causes are given below.

- Pollution, a gift of the industrial revolution can be given the pride of place for driving a variety of species in air, water and land towards extinction.
- Motorcars, air-conditioners and refrigerators, the three symbols of a modern, affluent society, have been instrumental in global warming and ozone

depletion. They have drastically altered the climate with disastrous effects on the various species.

- Indiscriminate use of toxic chemicals and pesticides and overexploitation of wildlife resources for commercial purposes are responsible for the rapid decline in the number of some species.
- Genetic erosion arises from the loss (due to commercial and anthropogenic pressures) of habitats rich in biodiversity and from the disappearance of the traditional conservation practices of wild species in their habitats by rural and tribal people.

Biodiversity conservation is done in *in-situ* or *ex-situ* methods. *In-situ* methods include protected areas like National Parks, Sanctuaries, Biosphere reserves and sacred forests and lakes. *Ex-situ* methods include seed banks, captive breeding, animal translocations, tissue culture banks, cryopreservation of gametes and embryos. Botanical gardens and zoological parks.

Conservation of Agro-biodiversity

Agricultural biodiversity is the outcome of the interactions among genetic resources, the environment and the management systems, and practices used by farmers. Mono cropping, crop intensification, disruption of soil and water regime are some of the factors for loss of agricultural bio-diversity. It is felt that changes in attitude and policy are needed to give practical effect to biodiversity conservation in agriculture. With the requirement to produce more food to meet the demand of growing population, single-species crops replaced mixed-species crops, and cropping on the same land was repeated at short intervals. In the process, soil fertility was depleted and populations of pest and disease species built up, being more readily transferred through contiguous plantings of the same cultivar. Organic fertilisers have long been used to address declining fertility. Pests and disease were initially controlled through crop rotation. During the 19th century as chemical knowledge increased and the technology to deliver chemicals to crops and soils improved, a different approach to agriculture became entrenched. Agricultural practice has changed to the extent that it has become a major factor in the destruction of natural ecosystems and the elimination of species. In the 20th century establishment of uniform fields were encouraged by mechanisation and the management cost savings of monocultures. Misused or overused pesticides and fertilisers had adverse effect on soil and water. Soils have been compacted by modern agricultural machinery during tillage operations. This restricts water penetration and increases the amount of surface water, causing runoff to carry away greater volumes of soil. Tree removal to open up semi-arid land for grazing and cropping has resulted in subsurface salt rising to the surface. The principal pollinators like honey bees of fruit trees and major staple crops perform environmental service. The major contributors to this decline in pollinator populations are fragmentation of the habitat of these species, agricultural and industrial chemicals that reduce their fertility and numbers, parasites and diseases, and the introduction of invasive alien species.

Agricultural productivity and protection of the biodiversity that underpins it can be achieved only by extending protected area concepts throughout the whole agricultural landscape. Mixed cultivation reduces the risk of economic loss caused by variations in weather conditions or from pest attack. In this method focus is on maximizing food security rather than on maximising yield. Maintenance and use of crop genetic diversity is part of farmers' coping strategies to deal with weather unpredictability and to shorten periods of food shortage by spreading food availability over time. Preservation of indigenous varieties through seed banks and their multiplication have proved to be an effective method of agricultural bio-diversity conservation.

The key to soil fertility is soil biodiversity. Microorganisms are major part of key ecological processes that sustain the functioning of ecosystems, including a range of associations between plants and microorganisms that efficiently fix atmospheric nitrogen through legume-Rhizobium associations, and extract phosphorus, various micronutrients, and water under very low moisture conditions, and through mycorrhizal associations. Organisms like Blue-green algae have an essential role in wetland rice agriculture. Maintenance of conditions that favour the co-evolution of pests and predators is an essential element of agro-biodiversity conservation.

Genetic biodiversity conservation

Pressure to increase crop and livestock yields leads to the development of new plant varieties and new animal strains (additions to diversity) through conventional breeding techniques and biotechnology. In traditional agriculture crop genetic resources are passed from generation to generation of farmers and are subject to a range of natural and human selection pressures. Environmental, biological, cultural and socio-economic factors all influence a farmer's decision to select or maintain a particular crop cultivar at any given time. Farmers, in turn, make decisions about planting, managing, harvesting and processing their crops that affect genetic diversity. Farmers make decisions on how much of each crop variety to plant each year, the percentage of seed or germplasm to save from their own stock and the percentage to buy or exchange from other sources. There are growing pressures on small farmers, who maintain this crop genetic diversity in the form of local cultivars.

Biodiversity International has led the development of methodologies for on farm conservation, and promoted the drafting of policies and strategies for the *in situ* conservation of crop wild relatives and their management inside and outside protected areas.

Fig. 8.4. Conservation of agriculture bio-diversity

8.7 Challenges to sustainable NRM

Sustainable use of natural resources is a challenge due to water scarcity and fragile nature of ecosystems including increasing pressure of burgeoning population on drylands. Consideration of these limitations is a first step for a policy framework that will lead to the sustained uses of natural resources while controlling the process of environmental degradation of these ecosystems.

Water scarcity: Availability of water and the stability of available water resources are crucial for the control of land degradation and combating desertification. Because agriculture is the major user of freshwater resources worldwide, intensified efforts need to be directed toward reducing the agricultural uses of water (Thomas et al., 1993; Gregersen et al., 2007). Providing reliable supplies of water for uses other than agriculture, therefore need be essential to people's well-being. Ensuring secured water supply and control of pollution require technical, institutional and political solutions.

Continuing land degradation and desertification: Increasing environmental degradation and desertification of dryland ecosystems is a problem all over the world. Studies have found that moderate to extreme soil degradation is caused mainly by inappropriate agricultural practices, on-going deforestation activities, and overgrazing of livestock by the early 1990s (World Resources Institute, 1992). This degradation represented 1.2 billion hectares or almost 12 per cent of the earth surface. Soil degradation in Africa and Asia is also caused by nutrient losses on land that is used for low-input agriculture and the effects of salinization resulting from poor management of irrigation systems (Chandra & Bhatia, 2000).

Continuing losses of limited soil resources by soil erosion are also widespread on throughout the dryland regions of the world. The continuing loss of forests, woodlands, and rangelands also contributes to environmental degradation and desertification processes of already fragile watershed landscapes as the hydrologic functions of these landscapes are diminished or even destroyed. The resultant losses of these ecosystems can lead to downstream flooding and the transport of excessive sediment loads and other pollutants into reservoirs. These impacts can be felt within a river basin, throughout a country, and even in neighbouring countries sharing a common river basin (Sharma, 1992; Brooks et al., 2003; Gregersen et al., 2007). Environmental degradation and desertification are likely to take place in ecosystems that are less able to recover from environmental stresses than other temperate and moist tropical ecosystems.

Socio-economic and demographic changes: Sustainable use of natural resources in drylands requires thorough understanding of the socio-economic issues. People living in dryland regions face several risks to their well-being due to the harsh environment conditions which affects their income generation and employment opportunities. Most of the inhabitants depend upon traditional subsistence farming and livestock raising.

8.8 Integrated Natural Resource Management (INRM)

Integrated Natural Resource Management is a process of managing natural resources in a systematic way, which includes multiple aspects of natural resource use (biophysical, socio-political, and economic) to meet production goals of producers and other direct users (e.g. food security, profitability, risk aversion) as well as goals of the wider community (e.g. poverty alleviation, welfare of future generations, environmental conservation). The conceptual basis of NRM has evolved in recent years through the convergence of research in diverse areas such as sustainable land use, participatory planning, integrated watershed management and adaptive management.

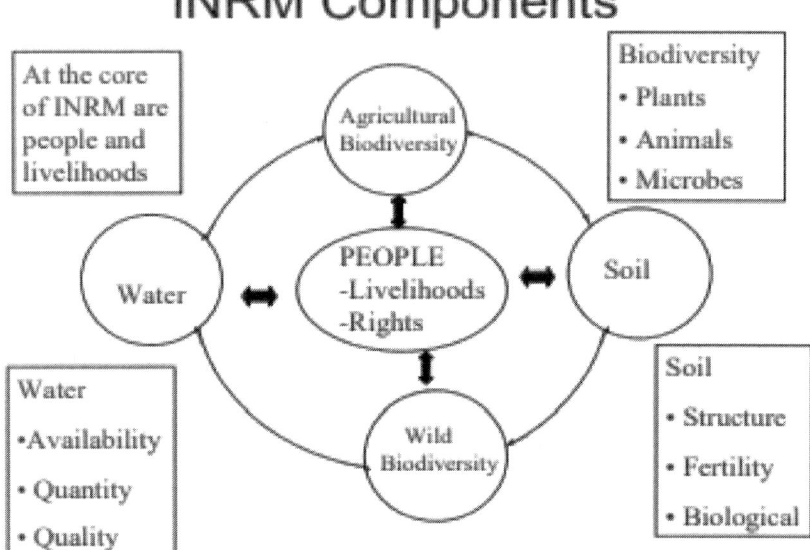

Fig. 8.5. Components of Integrated NRM

8.9 Community Natural Resource Management (CNRM)

Natural resources play a critical role in ensuring food security and livelihoods of communities. Sustainable management of these resources is critical to sustaining and improving the livelihoods of rural communities and from climate change point of view. It is well known that locals are better placed to conserve natural resources and people will conserve a resource only if benefits exceed the costs of conservation, Therefore, CNRM assumes great importance in rural areas for livelihoods enhancement of the communities. Participatory forest management has become very successful in many countries. CNRM requires combined effort of government, community and non-government organisations.

8.10 Sustainable use of natural resources

Agro-ecosystems cannot be sustainable in the long run without the knowledge, technical competence, and skilled labour needed to manage them effectively. In sustainable agriculture, the goal is to reduce the input of external energy and to substitute non-renewable energy sources with renewable sources (e.g., solar and wind power, biofuels from agricultural waste, or, where economically feasible, animal or human labour). Maintaining a high degree of genetic diversity by conserving as many crop varieties and animal breeds as possible will also provide more genetic resources for breeding resistance to diseases and pests.

Without improved efficiency measures, agricultural water consumption is expected to rise by about 20% globally by 2050. Climate change is already affecting water supply and agriculture through changes in the seasonal timing of rainfall and snow pack melt, as well as with higher occurrence and severity of droughts and floods. Hence sustainable use of water in agricultural production is very important.

Sustainable soil management is of particular importance in the context of climate change mitigation and adaption in the agricultural sector. Management of agricultural soils is of much importance in efforts by the agricultural sector to contribute to climate stability and meet the challenge of adapting to climate change. From farmers' perspective, soil management practices often lead to improved nutrient management and reduced need for inputs (fertilisers, machinery) thus resulting in cost savings, improved resource efficiency, and potentially greater economic stability. From a mitigation perspective, soil management practices at farm level play a role by preserving existing carbon content in agricultural soils. Practices such as the maintenance of grasslands, avoidance of drainage of organic rich soils such as peat soils, or rewetting of organic rich soils lead to important reductions of carbon emissions from land use. Soil management can increase soil organic matter content in mineral and organic soils, leading to carbon sequestration.

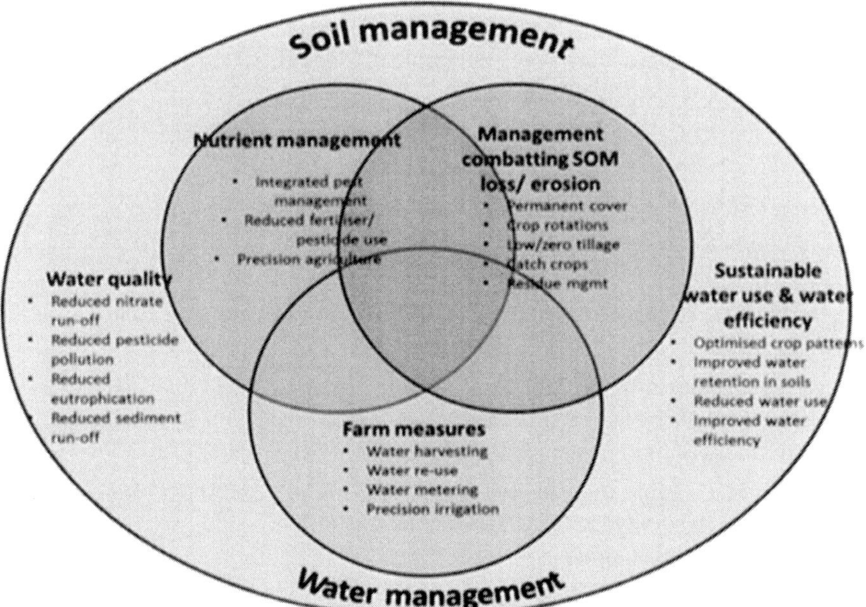

Fig. 8.6. Potential sustainable soil and water outcomes through improved land and farm management

Soil management can contribute to adaptation by improving soil quality and structure and thus enabling soils to deliver many vital ecosystem services which in turn contribute to the resistance and resilience of agricultural ecosystems to climate change by:

- Preserving the productive capacity (fertility) of soils;
- Improving the infiltration capacity and reduced run-off;
- Reducing soil erosion and nutrient runoff;
- Improving water holding capacity of soils (e.g., through addition of soil organic matter which is especially beneficial on sandy and clay soils);
- Reducing vulnerability of crops under drought conditions;
- Reducing pest / pathogen risks to crops; and
- Slowing down water runoff and reducing flood risk.

References

Anonymous, (2000): Natural Resource management for agricultural production in India (J.S.P. Yadav and G.B. Singh Eds)

Brooks, K. N. , P. F. Ffolliott, H. M. Gregersen & L. F. DeBano. 2003. Hydrology and the Management of Watersheds, Iowa State Press, ISBN 0-8138-2985-2, Ames, Iowa, USA

Chandra, S. & K. K. S. Bhatia. 2000. Water and Watershed Management in India: Policy Issues and Priority Areas for Research. In Land Stewardship in the 21st Century: The Contributions of Watershed Management, Ffolliott, P. F. , M. B. Baker, Jr. , C. B.

Fabricius, C., Koch, E., Magome, H., & Rurner, S. (Eds.). (2004). Rights, Resources and Rural Development: Community-Based Natural Resource Management in Southern Africa. London: Earthscan

Gregersen, H. M. , P. F. Ffolliott & K. N. Brooks. 2007. Integrated Watershed Management: Connecting People to Their Land and Water. CAB International, ISBN 978-1-84593-281- 7, Oxfordshire, United Kingdom.

Sharma, N. P. , editor. 1992. Managing the World's Forests. Kendall Hunt Publishing Company, ISBN 0-8403-7885-8, Dubuque, Iowa, USA.

Smyth, A.J. and Dumanski, J. 1993. FESLM: An international framework for evaluating sustainable land management. A discussion paper. World Soil Resource Report 73. Food and Agriculture Organization, Rome, Italy. 74 pp

Thomas, R. , M. Colby, R. English, W. John, B. Rassas & P. Reiss. 1993. Water Resources Policy and Planning: Toward Environmental Sustainability. U. S. Agency for International Development, Irrigation Support Project for Asia and the Near East, Washington, DC, USA.

World Resources Institute. 1992. World Resources 1992-93. Oxford University Press, ISBN 0- 19-506230-2, New York, New York, USA.

9

Nutrition Sensitive Food System

9.1 Food and nutrition security

Food security is achieved when adequate food is available and accessible and satisfactorily used and utilized by all individuals to live a healthy and active life. But food and nutrition security is achieved when people have physical, economic and social access to food of sufficient quantity in terms of variety, diversity, nutrient content and safety to meet their dietary needs and food preferences for a healthy and active life coupled with sanitary environment, adequate health and education care. In order to achieve food and nutrition security there is a need for transitioning toward nutrition sensitive food system.

9.2 Nutrition sensitive food system

Food systems are involved in the production, distribution, and consumption of food. A nutrition-sensitive approach not only considers policies related to macro-level availability of and access to nutrient-dense food, but it also focuses on household- and individual-level determinants of improved nutrition. Making the whole food system more nutrition sensitive requires a deliberate policy-oriented approach, which includes a combination of nutrition-specific as well as nutrition-sensitive interventions.

More diverse diets are balanced in calorie, protein, and micronutrient intakes that lead to better anthropometric outcomes for all age groups and better overall cognitive outcomes. As agriculture forms a major part of the food system, its role in enhancing nutrition always takes centre stage.

Meeting the demand for staple grains has become the primary challenge. With the rise of income level of people increased demand for diet diversification led to the rising demand for non-staple foods, such as vegetables, fruit, livestock, and dairy products. However, weak supply responsiveness for non-staples and their high relative prices, due to the persistence of policies that favour staple grains, resulted in limited access for the poor to a more nutritious diet which has caused malnutrition and under-nutrition. Therefore, policy focus on enhancing the diversity of the food system, particularly for micronutrient-rich horticultural and livestock products, is extremely crucial in these contexts.

Nutrition-specific interventions or programmes are those that address the *immediate determinants* of fatal and child nutrition and development-adequate food and nutrient intake, feeding, caregiving and parenting practices, and low burden of infectious diseases.

Nutrition-sensitive interventions or programmes are those that address the *underlying determinants* of fatal and child nutrition and development-food security; adequate caregiving resources at the maternal, household and community levels.

The food system is not limited to agricultural production, it involves intra-household equity, behaviour change, food safety, and access to clean water and sanitation. Re-shaping the current food system to be more nutrition sensitive, a deliberate policy-oriented approach, which includes a combination of nutrition-specific interventions, infrastructure investments, and producer incentives that complement each other are required (Pingali and Sunder, 2017).

To reduce malnutrition, a food systems perspective should be applied since a food systems approach seeks to address problems related to food insecurity and malnutrition by understanding the production requirements for diverse food to increase their supply, affordability, and consumption (Townsend 2015; Global Panel on Agriculture and Food Systems for Nutrition 2016; Per Pinstrup-Andersen 2012b; Kataki and Babu, 2002). Food systems are complex networks of individuals and institutions that provide food for everyone (Pinstrup-Andersen 2012a, FAO 2013). The food system of a country encompasses the different ways that food is produced, processed, distributed within the country and outside it, and consumed. The food system is governed by government policies and regulations, supported by various institutions and underpinned by public investments in infrastructure and private investments by food producers and distributors (FAO 2016).

Nutrition-sensitive agriculture and food systems contribute to improving health outcomes, through for example, production of diverse, safe and nutrient-rich food, income generation that can facilitate access to health services, through reducing contamination of water sources, and through the application of labour-saving technologies (FAO, 2016). To build measures to reduce malnutrition from food systems perspective, understanding relevant drivers of food security with an emphasis on nutrition is integral to the planning process (ADB, 2016; Kataki and Babu, 2003).

9.3 Nutrition sensitive agriculture (NSA)

Agriculture shares a symbiotic relationship with nutrition and health. There are multiple ways of conceptualizing the various pathways between agriculture and nutrition, and they all share many common features. Nutrition-sensitive agriculture is an approach that seeks to ensure the production of a variety of affordable, nutritious, culturally appropriate and safe foods in adequate quantity and quality to meet the dietary requirements of populations in a sustainable manner (FAO, 2017). Recently, the key recommendations were broadened to include improving nutrition through agriculture and food systems (Ag2Nut, 2014). Investment in nutrition can yield high returns in terms of reduced health costs, increased productivity and improved human resources capacity and economic growth (Covic & and Hendriks 2016, Shekar et al., 2017)

Nutrition sensitive agriculture can be implemented in areas like (i) making food more available and accessible, (ii) making food more diverse and production more sustainable through diversification, and (iii) making food more nutritious by bio-fortification. Bio-fortified staples, such as zinc-fortified rice and wheat, can play an important role in environments where markets are not well developed and where rural populations largely consume what they produce. Bio-fortified non-staples, such as pulses, could be an effective source of protein and micronutrients in areas with poor or better market infrastructure conditions.

Objectives

- Making food more available and accessible
- Making food more diverse and production more sustainable
- Making food more nutritious

Pathways to nutrition sensitive agriculture

There are six pathways through which agricultural interventions can improve nutrition (Ruel and Alderman, 2013). These are: (i) food access from own-production; (ii) income from the sale of commodities produced; (iii) food prices from changes in supply and demand; (iv) women's social status and empowerment through increased access to and control over resources; (v) women's time through participation in agriculture, which can be either positive or negative for their own nutrition and that of their children; and (vi) women's health and nutrition through engagement in agriculture.

9.4 Approaches to NSA

Homestead food production systems, home vegetable gardens, bio-fortified crops, small animals, livestock, fisheries, dairy etc. are included in NSA. Promotion of production diversity, micronutrient-rich crops (including bio-fortified crops), dairy, or small animal rearing can improve the production and consumption of targeted commodities.

9.4.1 Farming system for nutrition (FSN)

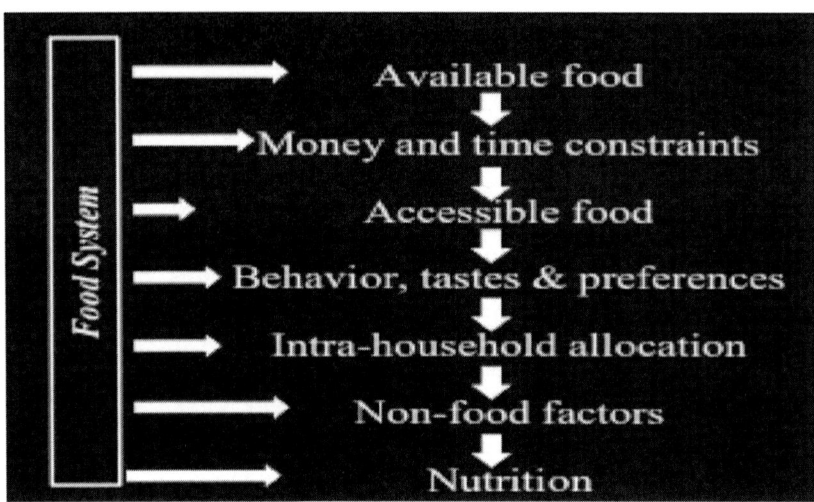

Fig. 9.1. Genetic food system to nutrition pathway

The FNS includes cultivation of nutritious millets, pulses, vegetables in nutrition garden and production of animal proteins from backyard poultry. This is schematically presented in Fig. 9.2.

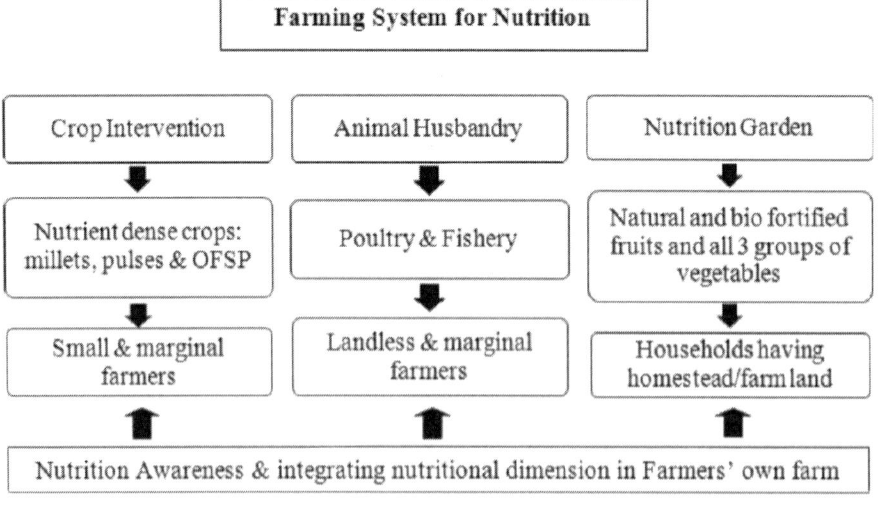

Fig. 9.2. Farming System to Nutrition

9.4.2 Millets as nutri-cereals

In the shifting cultivation (slash and burn system) farmers are accustomed of growing a series of millets along with pigeon pea (arhar) and rice to ensure food security as these crops give the harvest right from August-September to February-March in sequence. There is no doubt that the millets, coming under C-4 category of crops have enough production potential than rice and wheat and they are more nutritious than these crops. Millets require less water and are more resilient to climate vulnerability.

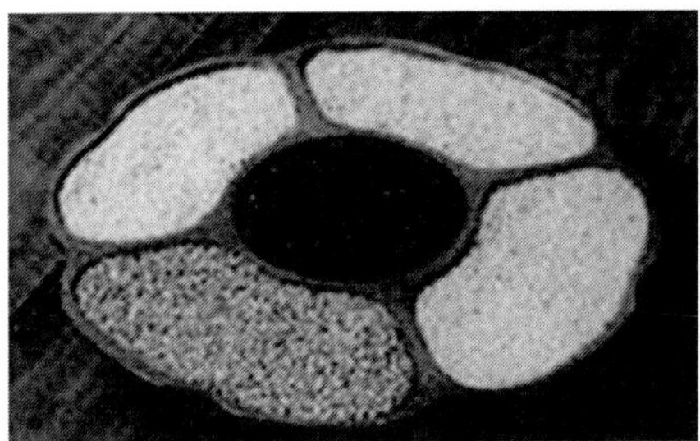

Fig. 9.3. Nutritious millets (Ragi, little millets etc.)

Millets such as finger millet (ragi), pearl millet (bajra) and small millets are a rich source of protein, vitamin B, and minerals such as magnesium, potassium, zinc, copper, and manganese, and contain significant amounts of phenol, which act as anti-oxidants and as preventive for degenerative illnesses such as heart diseases and cancer. People in tribal area say "Rice fills the stomach but does not last, and we are hungry and tired sooner. We must eat ragi (finger millet), which provides the energy to walk our steep hills every day." A 100 g pearl millet breakfast has 100% of daily iron requirement for adults while 100 g of finger millet has half the requirement of iron and one-third of calcium, important for children and lactating mothers.

9.4.3 Bio-fortification

There are two routes to address malnourishment and malnutrition in women and children. One is nutrition specific and the other is nutrition sensitive. Nutrition sensitive agriculture is the modern concept to ensure food and nutrition security through crop diversification, diversified food production, nutrition gardens, nutrition dense crops and bio-fortification.

Bio-fortification, the breeding of staple crops that are richer in essential micronutrients than traditional varieties, has been shown to be a feasible and cost-effective approach to addressing deficiencies in vitamin A, iron, and zinc (Bouis and Saltzman, 2017). Bio-fortification consists in developing new varieties of staple crops (i.e. cassava, maize, orange-fleshed sweet potatoes, wheat, rice, pearl millet, sorghum, banana, plantain, squash, beans, lentils and cowpeas) with the explicit intent of enhancing levels of bioavailable micronutrients (i.e. pro-vitamin A, iron and zinc). While bio-fortification is most commonly accomplished using conventional plant breeding, agronomic bio-fortification (i.e. application of micronutrient-rich fertilizers via soil or leaves) and transgenic techniques are also used. Bio-fortification programmes aim to increase micronutrient availability in staple crops themselves. Bio-fortification may also be a better choice than industrial fortification in rural areas where access to markets and commercially fortified products is poor. Bio-fortification of pulses and millets with iron, bio-fortification of maize, rice and wheat with zinc and bio-fortification of cassava, sweet potato etc. with carotenoid are at different stages of development.

The biological mechanism for bio-fortified crops improving nutritional status is simple: bio-fortified crops are more nutrient-dense than non-biofortified varieties. Therefore, assuming similar micronutrient bioavailability and retention after cooking or processing and storage, persons will consume and absorb more micronutrients from eating bio-fortified crops than from the same amount of non-biofortified crops. In populations with a diet limited in these micronutrients, the consumption of bio-fortified staple crops can improve micronutrient intake.

There are two main behavioural issues related to bio-fortification: one for the farmer and one for the consumer. Farmers are interested in planting new varieties that are agronomically superior to the current varieties they plant, for example cultivars that are more drought resistant, have more yield or less susceptibility to diseases. Therefore, bio-fortified varieties must be agronomically equivalent or preferably superior to the less nutrient dense market and traditional varieties with which they will compete.

Bio-fortified crops are often assumed to be transgenic or genetically modified crops. While the thrust of bio-fortification activities through international projects focused on the production of nutritionally enhanced crops through conventional breeding, increasingly, more efforts are using genetic modification to add value to staple and non-staple crops.

OUAT, Bhubaneswar is going to release 9 rice varieties rich in zinc and iron. Such varieties are OR2418-2, OR 2413-9 and OR2329-8 etc.CR Dhan 310 contains 10.3% protein. DRR Dhan 45 contains 25.2 ppm zinc. WB02 wheat is rich in zinc and iron. Pusa Vivek maize contains high pro-vitamin A and tryptophan with lysine. HHB 299 pearl millet contains high iron and zinc. Bhu Sona sweet potato contains high B-carotene and Bhu Krishna sweet potato contains high anthocyanin. ICAR has developed bio-fortified varieties like CR Dhan 310 (protein rich variety), WB 02 (zinc & iron rich variety) wheat, Pusa Vivek

QPM9 maize, Improved (provitamin-A, lysine & tryptophan rich hybrid), HHB 299 (iron & zinc rich hybrid) pearl millet , Pusa Mustard 30 (low erucic acid variety) and Bhu Sona (â-carotene rich variety) sweet potato.

The biofortified varieties are 1.5 to 3.0 times more nutritious than the traditional varieties. The rice variety CR DHAN 315 has excess zinc; the wheat variety HD 3298 is enriched with protein and iron while DBW 303 and DDW 48 are rich in protein and iron.

9.4.4 Pulses as a protein supplement

Pulses or grain legumes are jewel crops in India. These crops constitute an essential part of Indian diet because nearly 43% of all Indians are vegetarian (urban-48% rural-41%) and pulses contain 20-28% of protein with good digestive value. Endowed with nitrogen fixing bacteria in roots pulses fix appreciable amount of nitrogen (50- 200 kg/ha) and thus enrich soil fertility. Due to their shorter duration they can fit into any cropping system. Importance of pulses in maintaining food security as well as nutritional security demands that production of pulses definitely needs to be increased manifold to meet the demand in coming years.

Fig. 9.4. Pulse grains

In India pulses are grown in an area of 22-23 million hectares with an annual production of 20.26 million tons as average of five years. The production in 2019-20 was estimated as 23.02 Mt. India accounts for 33% of the world area and 22% of the world production of pulses. To meet the domestic demand and save foreign exchange India needs to produce 26.5 Mt per annum.

Pulse production in Odisha has shown tardy progress during last 25 years. During 1990-91 it produced 11.7 lakh tons from 21 lakh hectares with yield rate of 614 kg/ha. During 2019-20, 10.5 lakh tons are produced from 19.22 lakh ha with yield rate of 544 kg/ha against the national average of 806 kg/ha.

Fig. 9.5. Pulse crops (green gram, chick pea, pea)

Constraints in pulse production

Pulses are still considered as sick segment of our agriculture as they are cultivated under neglected condition. The promotional efforts through NFSM, RKVY and A3 P have shown some success in selected areas, but it is not enough to give a boost to pulse production to meet the demand. Recently, GoI has made provisions of about Rs.500 crores for promotion of pulses by way of enhancing breeder (ICAR and NARES) and quality seed production with a provision to establish 150 Seed Hubs and large scale demonstrations through KVKs. "Pradhan Mantri Krishi Sinchai Yojana" and "Pradhan Mantri Fasal Bima Yojana" of the Government of India can be a game changer in favour of pulses growers if micro-irrigation and farm input supply can be linked. With the availability of technologies including high yielding varieties developed by ICAR and NARS and implementation of appropriate policies like ensured supply of quality seed and other farm inputs, remunerative MSP, procurement, investment in creation of required infrastructure for seed/grain buffer stocks, it is possible to make a breakthrough in pulse production. Recently GOI has hiked MSP of pulses at an equivalent price of 150% of cost of production, but procurement under MSP is still a problem. Government can supply a small quantity of pulses along with rice under NFSA for nutrition security of target beneficiaries.

OUAT has identified poor soil, poor crop stand, poor crop management, poor nutrient management and poor technology along with seed constraint and moisture stress as predicaments in pulse production.

Future strategy

A twin approach of horizontal expansion and increasing productivity is possible. Although some 500 varieties of pulses have been released in India there is dearth of cold tolerant mung and Biri varieties for Odisha. Pigeon pea hybrid is a notable contribution of research. Similarly large seeded lentil IPL 406, green seeded pea IPFD 10-12, and extra-large seeded chick pea MNK1, PKV Kabuli 4-1 etc. can make a breakthrough in pulse production. Integrated crop management technologies have been standardised in forms of nutrient management, crop establishment, weed management, integrated pest management, and post- harvest management. Additional area could be brought under sequence cropping, intercepting, *paira* cropping (under sown) and crop substitution. Income of farmers could be raised by cultivating large seeded and green varieties, hybrids and mechanisation. There is a need for policy support like technology transfer, value addition, development of storage infrastructure and marketing through MSP route.

There are many alternatives like incentives to pulses growers as a cost for contribution of pulses in improving soil health for sustaining agricultural production base, improving total factor productivity of pulses like wheat and rice . We should break the jinx that productivity of pulses in Odisha is around 500 kg per ha during last 50 years. We should be ashamed for the treatment given to the jewel crop.

OUAT has suggested the strategies in form of better crop planning, improving seed replacement rate from 7% to 10%, participatory seed production, production of rhizobium culture and bio-agents, cluster demonstration of better technologies, capacity building and strong research to accelerate the production and productivity. Pulse production in Odisha has been enhanced by correction of soil acidity and molybdenum treatment.

9.4.5 Oilseeds

India's oilseeds production was a mere 5.26 million ton in 1949-50 which has jumped to whooping 32.10 million tons in 2016-17.The per capita consumption of oil was 3 kg per person in a year in 1950 which has increased to 18.50 kg in 2016-17 against world average of 25 kg and WHO's recommendation of 11 kg per person in a year. In spite of many schemes after Yellow Revolution, India imports about 14.01 million tons of edible oil and spends 73048 Cr rupees for this in the year 2016-17. GoI has estimated that India could produce 45.64 million tons of oilseeds from 31.20 million hectare in 2022-23.India can produce 17.03 million tons of edible oil from primary sources (9 oilseeds crops) and secondary sources (cotton seed, oil palm and tree crops)

and still will have deficit of 16.13 million tons for import by spending more than 80000 crore INR..

India grows 26.67 million ha of nine oilseed crops and produces 30.06 million tons (average of three years ending 2016-17) with 1130 kg per ha. It is estimated to grow nine oilseed crops in 31.20 million ha to produce 45.64 million tons in 2022-23.This will produce 11.41 million tons of oil and oil palm and cotton seed will produce 5.62 million tons making a total of 17.03 million tons of oil. Of the 26 million ha of oil crops, 34% is covered by soya bean followed by 27% each in groundnut and rapeseed-mustard, making a total of 88% which produce more than 80% of vegetable oil. Mustard produces 35% followed by 25% from groundnut and 21% by soya bean.

Fig. 9.6. Oilseed crops (groundnut, mustard, sunflower, soybean)

Constraints in production

The growth is negative in all oilseed crops except castor, sesame, mustard and soya bean. There is a yield gap of 51%, ranging from 21% to 149% between crops. Due to dominance of staple cereals, these crops are cultivated in mostly unirrigated areas up to 85% and on sub-marginal lands. Seed quality, management practices are often sub-optimal and there is heavy load of pest attack and post-harvest losses. Oil extraction is poor and the crops have been unremunerative due to marketing problems. There is nutrient deficiency in all crops with regard to NPK and zinc, boron, sulphur and molybdenum. The crops are affected by moisture stress at critical stages. Soil acidity reduces crop yields in ground nut and soya bean.

Government of India is implementing specific schemes like ISOPOM in 10th and 11th FYP, NMOOP in 12th Plan and NFSM (OS&OP) in 13th Plan. But the annual fund release is getting reduced (713.42 Cr in 2010-11 to 558.14 Cr in 2013-14, 327.50 Cr in 2016.17 and 265.50 Cr in 2017-18).There is over consumption of edible oil as the current use is 18.5 kg per person in a year against WHO's recommendation of 11 kg.

Future strategy

Oilseed crops can be given a better treatment by way of crop diversification and crop rotation in irrigated and fertile lands. The yield gap of 51% can be abridged by improvement in seed quality, better agronomy, moisture management, integrated pest and nutrient management. SAUs, ICRISAT, Directorate of Oilseeds Research and Regional Research Stations have evolved a wide array of new high yielding varieties and hybrids of oilseeds. These varieties could be multiplied in seed production chain and supplied adequately to the farmers before the cropping season. Dehumidified cold storage for groundnut can be installed. New varieties can be given to oilseed growers in form of composite seed kits along with Rhizobium culture (for ground nut and soya bean) and micro-nutrients. The oil recovery could be increased by installing mini oil mills in strategic locations. We need to incentivise the cultivation of oilseeds on a per hectare basis. Public awareness is required to reduce the consumption of edible oil (preferably saturated fat) limiting to 30 g per adult equivalent daily or 11 kg per person per year. The R&D efforts and field extension should be further strengthened by more allocation of funds and deployment of quality manpower.

9.4.6 Nutrition garden

Food meets the caloric requirement of a person, while nutrition is required for health. Satisfying the hunger or taste doesn't necessarily mean meeting the nutritional requirements of the body. Nutrition is essential for meeting the nutrient requirement of the body. Good nutrition is the foundation of good health. Well-nourished children grow and develop properly, are less sick less often, achieve more at school and grow up to be healthy and productive adults who in turn produce healthy children. Healthy people are the basis of successful and productive families, communities and societies. 'Malnutrition' implies to an overall and marked nutritional deficiency in the body. However, excess of nutrition can also cause health issues such as obesity. Malnutrition occurs when we do not get enough of the right kinds of food to maintain good health. Under-nutrition is lack of proper nutrition. It is caused by not having enough food containing nutrients required for growth and health. Under-nourished children do not grow or develop properly, suffer from diseases and cannot perform well at school. They become less productive adults and have fewer employment opportunities, meaning they earn less money. Undernourished women tend to give birth to underweight babies, perpetuating the cycle of poor nutrition in a community.

Balance diet is food which contains required quantity of carbohydrates, protein, fat, minerals and vitamins required for a good health. The concept of 'balanced diet' is promoted which is essentially based on the nutritional requirements that may vary individually according to age, gender, profession, and the like. A common balanced diet for Indians is a combination of rice/chapatti, dal, vegetables, fruits, milk, curd, egg or meat.

Fig. 9.7. Balance diet & food pyramid

Macro and micronutrients

- Macro-nutrients are carbohydrates (rice, wheat, sugar, roots and tubers), proteins (dal, egg, meat, milk, cheese), fats (edible oil, ghee, butter), dietary fibre (vegetables, whole grain and fruits). Carbohydrates supply instant energy; protein helps body building and repair of cells including producing enzymes and hormones; fats supply energy, supports cell growth, keeps the body warm and protects organs: and fibre protects from gastrointestinal problems, and even against type-2 diabetes.
- Micro-nutrients are minerals like iron, zinc, sodium, calcium, and potassium, etc. which help our body maintain the bone & tooth health, nerve & muscle function, and immune system, etc. We get micro-nutrients from nuts, beans & lentil, dark green leafy vegetables, milk (calcium), small fish etc.
- Vitamins (A, B, C,D, E, K etc.) are organic substances required by the body to maintain the normal process of metabolism. Their deficiency can cause various diseases (like night blindness due to deficiency of Vitamin-A). The important sources vary according to the vitamin, like citrus fruits for Vitamin-C and Cod Liver Oil for Vitamin-A.

Nutrient and food groups

All food contains important substances called nutrients which our bodies need in order to function properly. Nutrients help us to grow properly, give us energy to think, work and play and keep our bodies strong and healthy. Different foods contain different amounts of nutrients. We can group foods according to the amounts of different nutrients which they contain.

- White group – energy-giving foods
- Brown group – body-building foods
- Green group – protective foods
- Orange group – protective foods

Fig. 9.8. Food groups

Nutrition garden is an improved form of kitchen garden where selected vegetable/ fruit crops are grown systematically so as to meet the nutritional requirements of the family and sell the surplus in the market. Essentially it produces nutrition in the field.

Benefits of a nutrition garden

It is a source of fresh and nutritious vegetables for the family throughout the year. It helps ensure a quality control in the production so as to maintain the food & nutritional safety of the products. For instance, it is easy to go for a fully organic home garden. Availability of perennial crops like drumstick leaves can meet the requirement even at the odd hours of the day. By nutrition garden one can reduce the expenditure in buying vegetables and can make effective use of resources. What's more, working in the garden refreshes the mind and inspires a positive attitude.

Fig. 9.9. A family nutrition garden

Lay-out of nutrition garden

Sites selection is very crucial for success of a nutrition garden or kitchen garden. Usually it is established in backyard of house or on an open space, near water source receiving plenty of sunlight. Size and shape of garden depends on availability of land, number of persons in family and spare time available for its care. The size of a Nutrition garden depends upon the availability of land and number of persons for whom vegetables are to be provided. There is no restriction in the shape but as the houses are on hilly terrain, so suggested to go for terrace farming.

About 200 square m of land is sufficient to provide vegetables throughout year for a family consisting of five members. The area may be modified if the aim is to meet the family need and selling the surplus in the market.

General principles to be followed

- Shade loving vegetables like water leaf may be planted in perennial plot.
- Perennial vegetables like drumstick, curry leaf etc. should be allotted to one side of the garden so that they may neither shade remaining plants nor they interfere with intercultural operations.
- One or two compost pits may be provided on one corner of nutrition/ kitchen garden for effective utilization of farm / kitchen waste.

- Fences on all sides should be made with barbed wire or with live fence to protect from pilferage and .animal menace. Fence may be made strong by planting Agasthi (*Sesbania grandiflora*) at 1.0 m. distance along the fence.
- After allotting areas for perennial crops, remaining portion may be divided into 6-10 equal plots for raising annual vegetable crops. Crop rotation, mixed cropping and inter cropping have to be followed.
- Provide walking path at centre as well as along four sides for movement and agricultural operations.
- Utilise ridges, which separate individual plots for growing root and tuber crops.
- As intensive and continuous cropping is done in a nutrition/kitchen garden, fertility and texture of soil may be maintained by applying adequate quantities of organic manure frequently.
- Since fresh vegetables are directly utilised, follow clean cultivation, mechanical removal of pest / disease affected plants, planting of resistant varieties, biological control, use of bio-pesticides or bio-fungicides for pest and disease control in a kitchen garden.
- In a nutrition garden, preference should be given to long duration and steady yielding varieties than high yielding ones, which require constant care.
- While allotting or arranging crops in each sub-plot, care should be taken to plant varieties / crops at ideal time of planting or season.
- A bee-hive may be provided for ensuring adequate pollination of crops besides obtaining honey.

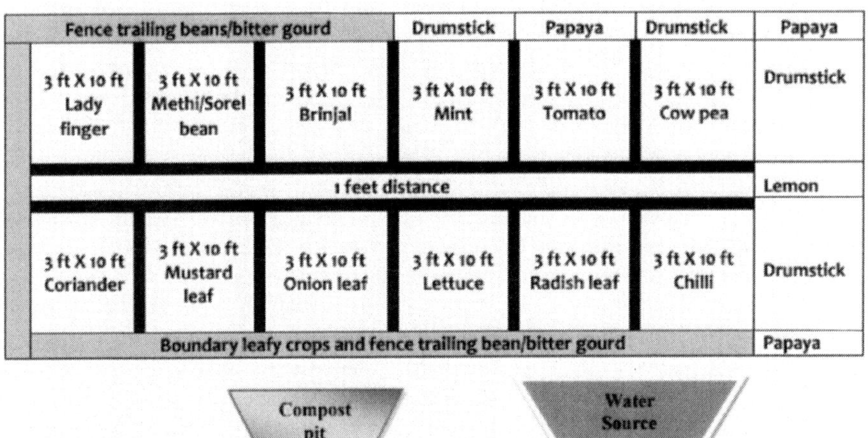

Fig. 9.10. Model Lay-out of a nutrition garden

Criteria for selection of fruits and vegetables for a nutrition garden

The family requires a mixed meal to provide balanced diet to all members. The mixed meal is a balanced combination of water, staple cereals, vegetables, fruits, legumes, sugar, oil, milk, fish/meat and egg.

Table 9.1. Family food requirement

Food item	What they supply
Cereals	Energy, protein, vitamin B & E
Roots and Tubers	Energy and vitamins
Legumes (dal)	Energy, protein, fat,. Vitamin E, calcium, iron and zinc
Oil and nuts	Energy, vitamin E and A
Vegetables and fruits	Minerals and vitamins for protection of body
Animal foods	Protein, minerals like iron, zinc, calcium and vitamin A & D
Water	Body contains 80% water which are required all metabolic activities and removal of wastes.

While we can get cereals, dal, oil, sugar and non-veg items from the market, we can get vegetables and fruits from our garden. The food groups have been shown in fig.9.8 in different colours. Such food groups are indicated in the following Table 9.2.

Table 9.2. Different food groups

Group	Examples	Main nutrient supplied and function
WHITE GROUP Energy-giving foods	Cereals (maize, sorghum, millet, wheat, rice), Roots and tubers (sweet potatoes, yams, collocasia, cassava, potatoes)	Good source of energy, protein, some vitamins and minerals. Unrefined cereals are also a good source of fibre. Minimises the risk of colon, prostate and breast cancer.
BROWN GROUP Body-building foods	Legumes and oilseeds (Cowpea, pigeon pea, beans, groundnuts, sunflower seeds, soybean, sesame)	Good sources of protein, some vitamins, minerals, fibre and oil
GREEN GROUP Protective foods	Dark green leafy vegetables (spinach, poi, amaranths, lettuce)	Excellent sources of vitamins, folic acid, minerals and fibre. Protect your body from cancer, cholesterol, regulate digestion and improve immune system
ORANGE GROUP Protective foods	Orange and yellow fruit and vegetables (pumpkin, sweet potato, carrot, turnip, radish, tomatoes, peppers, onions, gourds, green beans, mango, papaya, guava, banana)	Provide vitamins and antioxidants. Protect your nervous system and prevent heart diseases. They play a role in maintaining skin health, eye sight, boosting the immune system and bones.

It is desirable to select vegetables and fruits from the above four groups to meet the nutrition requirement of our family members. The following checklist may be used:

- Choose crops (vegetables/fruits) on the principles of good nutrition.
- Malnutrition occurs when we do not get enough of the right kinds of food to maintain good health
- The period from conception to 2 years of age (the first 1,000 days) is the most important time in a person's life for good nutrition.
- In order to stay healthy we must have food from each of the four sections of the colour wheel each day. Because different foods contain nutrients in different amounts we need to eat many different types of food to get all of the necessary nutrients.
- We cannot stay healthy by eating just one or two of the same foods each day. If we grow many different types of crops at home and at school we can add more variety to our diet.
- Family members who must take extra care to have a good diet include babies and children, women who are pregnant or breastfeeding and people who are sick.

Below are some issues which should be considered to decide what to grow in the nutrition garden:

- Select vegetables that are preferred and likely to be eaten in local diets
- Select 10-12 different leafy and green vegetables for each season (duration of 4 months) referring colour indicator chart
- Select some fruits for the North Western boundary (Papaya, Drumstick, Lemon)
- Choose some fence crops such as trailing bean, bitter gourd, cucumber etc.
- Select vegetables that are hardy, easy to grow and manage, and well adapted to the local climate and soil
- Select vegetables that produce good yields and are tolerant to common pests and diseases

What are the crops needed to be grown in our nutrition garden?

It may not be possible to all food items that are needed to grow in the nutrition garden. Therefore the bare necessities to supplement the family nutrition can be grown the garden. We can adopt the following guidelines before selecting the type of fruits and vegetables to be grown in our garden.

Selection of crops

It is important to grow all kinds of vegetables that the family members like and require depending upon the area and other facilities available. If land is limited, it is advisable to grow crops that produce higher yields per unit area and time. If land is sufficiently available, then crops such as banana, papaya, guava, and lemon can be grown in a nutrition/home garden. It is required to select the varieties of crops as per resource and environment (Table 9.3).

Table 9.3. Criteria for selection of crops

Criteria	Crop varieties to be selected
Environment and soil	For shady places: Turmeric ginger and different types of colocasia For partial shade: Spinach, elephant foot yam, ginger, turmeric, Use of high trees as support: Yam, flat beans, black pepper Use of trellis: pumpkin, pointed gourd, bitter gourd, ridge gourd, little gourd, snake gourd, cucumber etc. Use of marshy lands: Colocasia, Sunusunia
According to soil pH	5.5 to 7.0: Tomato, brinjal, chilli gourds, pumpkin, cucumber etc. 6.0 to 7.0: Spinach bean, etc. 6.0 to 7.5 : Bush beans, beet, radish, green peas, etc. 6.0 to 8.0 :Lady finger, sweet potato, etc.
Water requirement	Very Low: Water melon Pumpkin, Ash gourd Low: Cluster bean French bean, Cow-pea, Pea ,Ridge gourd, Bottle gourd Moderate: Tomato Chilli, Onion, Potato, Carrot, Cucumber High: Leafy vegetables Cabbage, Cauliflower Broccoli
According to rooting depth	Very shallow15-30 cm: Onion, Lettuce, Small radish Shallow rooted (30-60 cm) :Cabbage, Cauliflower, Celery, Garlic, Leek, Palak, Radish, Cow pea, Potato Moderately deep rooted (60-90 cm): Beet root, Brinjal, Cucumber, Musk melon, French Bean Deep rooted: (90-120 cm) Chilli, Turnip, Summer Squash Very deep rooted (120-180 cm: Water melon, Pumpkin, Winter Squash
According to growing season	Kharif (June- Oct): Cucurbits, Brinjal, Chilli, Okra, Tomato, Ginger, Cowpea, Cluster bean, Cucumber, Sweet Potato, etc. Rabi (Oct-Feb): Carrot, Beet root, Radish, Turnip, Cabbage, Cauliflower, Potato, Brinjal, Tomato, Chilli, Okra, Cow pea Summer(Feb-May): Brinjal, Tomato, Chilli, Okra, Cow pea, Cluster Bean, Cucurbits,
According to light requirement	Low light demanders: Leafy vegetables Medium light demanders: Potato, Radish, Turnip, Carrot, Palak, Beet root, Onion, Cabbage, (require at least half a day of sunlight) :Cauliflower Photo-intensive(require full sun) Tomato, Brinjal, Okra, Chilli, Capsicum, French bean, Cucumber, Cow Pea

Harvest period	Quick: Leafy vegetables Long time: Lemon, mango, guava Annual fruits: Banana, Papaya Minimum care: Drumstick
According to specific nutrition	Protein: Cow pea, Cluster bean, French bean, bean, Yard long bean Starch: Roots and tubers, Banana Minerals: Calcium: Beet, Amaranths, Fenugreek, Turnip, Coriander, Pumpkin, Onion, Tomato Potassium: Sweet potato, potato, Bitter gourd, banana Phosphorus: Garlic, Peas, Bitter Gourd Iron: Bitter gourd, Drumstick, Amaranths, Fenugreek, Spinach, poi Vitamin A:Carror, Broccoli, Squash, Sweet potato, Pumpkin, Cabbage, Tomato, Coriander, Broccoli Vitamin B: Citrus, Beans, Avocado, Garlic, Colocasia Vitamin C: Pepper, Strawberry, Broccoli, Potato, Lime, Turnip, Cauliflower, Bitter gourd, Radish Vitamin D:Spinach, Okra, Soybean Vitamin E: Pumpkin, Beet, Spinach Vitamin K: Leafy vegetables, Broccoli, Cauliflower, Cabbage

Crop rotation

Crop rotation is a practice of growing different crops in succession on a piece of land with specified period of time, with object to get maximum profit from least investment without harming the soil fertility. It is commonly used to control diseases and insect pests in the vegetable garden, and to build up the organic matter and soil nutrients that certain plants use during their life cycle. Different plants take different nutrients out of the soil and add back other elements or enhance the soil in other ways. Therefore, to prevent your garden from becoming less productive from season to season, crop rotation is highly recommended.

Crops can be selected as per their demand of nutrients:

- Those that demand heavy nutrients are known as heavy feeders, like leafy vegetables (e.g., spinach) and fruit vegetables such ash pumpkin, cucumber, and tomatoes etc., as well as grains like corn.
- Most roots and tubers, such as beets, carrots, yams and taros, and herbs & spices such as ginger, turmeric, chilli, garlic, and bunching onion, etc. are considered medium feeders as they remove less nitrogen.
- Most beans and peas are considered low feeders and can be grown on poor soils. They even add nitrogen to the soil.

Considerations for a good crop rotation

Crops from same family should not be grown together in the same plot to avoid diseases and pests. Some of the main families are: Cucurbitaceae (cucumber, pumpkin and all the gourds); most of these plants need water, but do not like poorly drained soils); Solanaceae (Brinjal, tomato, chilli, potatoes etc.); Fabaceae (beans and peas); Convolvulaceae (water spinach and sweet potatoes); Malvaceae (lady finger and roselle); and Chenopodiaceae (beet root and spinaches).

How to plan a rotation?

- List out vegetables to be planted in the garden. Divide the vegetables into four groups based on what part of the plant you plan to eat, i.e., plants grown for leaves or flowers (amaranth, lettuce, broccoli, cabbage), fruits (tomato, eggplant, pepper, cucumber), roots (carrot, radish, onion) and legumes (beans, peas, cover plants/green manure like alfalfa or clover).
- Group the plants together in botanical families. This is to help you understand what crops are closely-related. In general, crops in the same family should not be planted in the same field continuously. Members of the tomato, cabbage or pumpkin families should not be planted in the same place more than once every four years. Neither should members of the tomato family be grown in rotation with members of the pumpkin family, since they can pass on diseases to one another.
- Heavy feeder should be followed by a light feeder/medium feeders
- To avoid pests, we can try also to grow aromatic herbs (with strong smell), in-between or around the main crops. Some of these plants are basil, coriander, mint, small marigold, button chrysanthemum, and Indian borage, etc..
- There are specific intercrop combinations too for pest control, like Cabbage + Carrot against Diamond Black Moth, Okra + Cowpea against Yellow Vein Mosaic Virus, Tomato + Basil against hornworm, and Cucumber + Radish against cucumber beetles. Such strategies are in fact a part of integrated farming.
- Draw a map of the growing area and divide it into equal sections.
- Work out which crops will be planted in which area.
- Keep records of what actually happens, not just what you planned.
- The year of rotation should be 2-4 years.

- If you start by planting a leaf crop in a bed such as spinach, cabbage or amaranths ,the next crop to be planted in that bed should be a root crop such as onion, potato or carrot. After the root crop, plant a fruit crop (such as tomatoes, aubergine, pepper or chilli) and finally, plant legumes (peas or beans).

Sample crop rotation

Year	Plot-1	Plot-2	Plot-3	Plot-4
1	Legume	Leaves	Fruits	Roots
2	Leaf	Fruits	Roots	Legumes
3	Fruit	Roots	Legumes	Leaves
4	Root	Legumes	Leaves	Fruits

Selection of crops and varieties

The success of a nutrition garden depends upon proper selection of crops, varieties and cultivation practices.

Vegetables	Amaranth leaves, spinach, methi etc., Gourds(pumpkin, bitter gourd, small gourd, pointed gourd, ridge gourd, ash gourd, bottle gourd), roots and tubers (potato, sweet potato, beet, turnip, radish, carrot, yam, colocasia, tapioca, elephant foot yam), cole crops (knolkhol, cabbage, cauli flower, Broccoli), fruit vegetables (okra, capsicum, tomato, brinjal), tree vegetables (moringa, papaya, banana, jack fruit), Beans (French bean, dolichos bean), Sweet corn, Peas etc.
Fruits	Banana, Guava, pomegranate, pine apple, custard apple, lemon, orange, mango, dragon fruit etc.
Spices	Chilli, coriander, fenugreek, ginger, turmeric, black pepper, onion, garlic

Selection of varieties

The selection of varieties can be made in consultation with local Horticulture Staff. Some promising local varieties can also be selected. The list of some improved varieties is given in Table 9.4.

Table 9.4. Recommended varieties.

Crop	Variety
Tomato	Arka Rashkhak, Arka Samrat, Arka Abhed, BT, Hybrids
Okra	Arka Anamika, Arka Abhay, Arka Nikita
Brinjal	Arka Keshav, Arka Nilakanth, Arka Ananda
Cow pea	Arka Garima, Arka Suman, Arka Mangala
Bean	Arka Komal, Arka Subhida, Arka Anup, Arka Arjun
Chilli	Arka Meghana, Arka Haritha

Onion	Arka Kayan, Arka Niketan
Amarant	Arka Samraksha, Arka Arunima
Palak	Arka Anupama
Radish	Pusa Chetki, Pusa Rasmi, Japanese White
Bean	Contender, Yard long Bean, Runner bean, Kentuky wonder
Yam	Shree Kirthi, Shree Roopa, Shree Silpa, Shree Latha
Cowpea	Pusa Barsati, Pusa Komal, Pusa Dofasali
Sweet potato	Gouri, Shankar, Shree Bhadra, Shree Nandini
Drum stick	PKM-1, PKM-2
Potato	Kufri Jyoti, Kufri Pukhraj, Kufri Lalima
Cauli flower	Snow Ball, Pusa Deepali, Arka Kanti, Arka Vimal
Cabbage	Pride of India, Golden Acre, Pusa synthetic, Pusa Ageti, Hybrid
Knolkhol	Purple Viena, Pusa Virat, Early white Viena
Capsicum	California wonder, Chinese giant, Yellow wonder, Arka Mohini
Pumpkin	Arka Chandan, Guamal, CM-14, Pusa Viswas

Cultivation practices

Land preparation, fencing, manuring, appropriate sowing or planting after seed treatment, inter culture, mulching, nutrient management, water management and pest management are the important practices for a nutrition garden. It is better to adopt organic farming methods in a nutrition garden, which will keep the family members healthy.

References

ADB. (2013). Food Security in Asia and the Pacific. Asian Development Bank https://www.adb.org/sites/default/files/publication/30349/food-security-asia-pacific.pd

Ag2Nut CoP (Agriculture-Nutrition Community of Practice). Key recommendations for improving nutrition through agriculture and food systems. 2014. http://unscn.org/files/Agriculture-Nutrition-CoP/AgricultureNutrition_Key_recommendations.pdf. Accessed 29 Dec 2014

Bouis, H. 2014. Biofortification Progress Briefs August 2014. Washington DC, Harvest Plus, August 2014 www.harvestplus.org/sites/default/files/ bioofortification. Progress Briefs. August2014EB2.pdf

Bouis, H.E , Saltzman, A, 2017. Improving nutrition through biofortification: a review of evidence from harvesPlus, 2003 through 2016.

Covic, Namukolo, ed.; and Hendriks, Sheryl L., ed. 2016. Achieving a nutrition revolution for Africa: The road to healthier diets and optimal nutrition. ReSAKSS Annual Trends and Outlook Report 2015. Washington, D.C.: International Food Policy Research Institute (IFPRI). http://dx.doi.org/10.2499/978089629593

FAO. (2013). Overview of Nutrition Sensitive Food Systems: Policy Options and Knowledge Gaps

FAO. (2016). The State of Food and Agriculture: Climate Change, Agriculture, and Food Security. http://www.fao.org/3/a-i6030e.pdf

FAO. (2017). Nutrition-sensitive agriculture and food systems in practice: Options for intervention. http://www.fao.org/3/a-i7848e.pdf

HarvestPlus. 2016. Biofortification. The Evidence. www.harvestplus.org/node/609

Kataki, P., and S.C. Babu. (2003). Food Systems Approach to Nutrition. Haworth Press, New York

Kataki, P., and S.C. Babu. (2003). Food Systems Approach to Nutrition. Haworth Press, New York

Pingali, P., & Sunder, N. (2017). Transitioning Toward Nutrition Sensitive Food Systems in Developing Countries. Annual Review of Resource Economics. doi:https://doi.org/10.1146/annurev resource100516-053552

Pinstrup-Anderson, P. (2012a). Food Systems and Human Health and Nutrition: An Economic Policy Perspective with a Focus on Africa. The Center on Food Security and the Pinstrup-Anderson, P. (2012b). Policy Interventions to Promote Nutrition Sensitive Food Systems

Ruel, M.T., Alderman, H., 2013. Nutrition-sensitive interventions and programmes: how can they help to accelerate progress in improving maternal and child nutrition? Lancet 382, 536–551. http://dx.doi.org/10.1016/S0140-6736(13)60843-0.

Shekar, M., Kakietek, J., Eberwein, J., & Walters, D. (2017). An Investment Framework for Nutrition: Reaching Global targets for Stunting, anaemia, Breastfeeding and Wasting. Washington DC, United States of America: The World Bank Group

Townsend, Robert. 2015. Ending poverty and hunger by 2030: an agenda for the global food system (English). Washington, D.C.: World Bank Group. http://documents.worldbank.org/curated/en/700061468334490682/Ending-poverty-and-hungerby-2030-an-agenda-for-the-global-food-system

10

Conservation Agriculture

10.1 What is conservation agriculture (CA)?

Conservation Agriculture Conservation agriculture is an approach to managing agro-ecosystems, for improved and sustained productivity, increased profits and food security while preserving and enhancing the resource base and the environment. It is a set of technologies, including minimum disturbance, permanent soil cover, diversified crop rotations, and integrated weed management (Reicosky and Saxena, 2007; Hobbs et al., 2008, Friedrich et al., 2012). Conservation agriculture aims at reducing and/or reverting many negative effects of conventional farming practices such as soil erosion (Putte et al., 2010), soil organic matter decline, water loss, soil physical degradation, and fuel use (Baker et al., 2002, FAO, 2008). It enhances biodiversity and natural biological processes above and below the ground surface, which ultimately increases water and nutrient use efficiency.

The elements of conservation agriculture are minimum disturbance of soil, permanent soil cover, diversified crop rotation and weed control (Fig. 10.1).

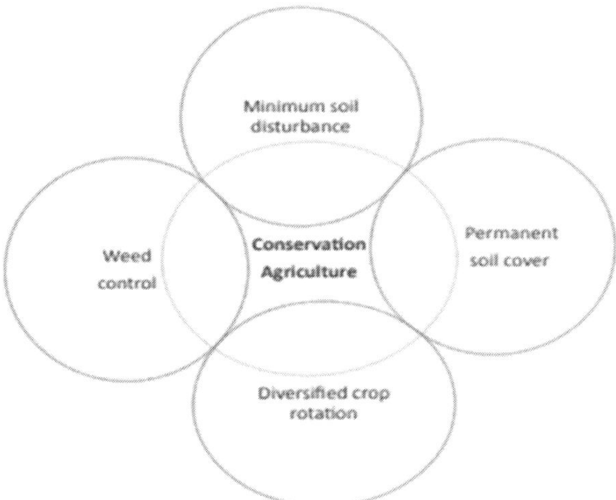

Fig. 10.1 Elements of conservation Agriculture (Farooq, M and Siddique, H.M)

10.2 Characteristics of CA

There are some similarities between organic farming and conservation agriculture. Both maintain a balance between agriculture and resources and protect the soil and soil organic matter content. However, the main difference between these two types of farming is that organic farmers use a plough or soil tillage, while conservation agriculture uses no-till system. Conservation agriculture and climate smart agriculture are also similar, but their purpose is different. While conservation agriculture aims at positive effect on environment using natural processes in spite of climate risks, climate-smart agriculture aims to adapt to and mitigate the effects of climate change by sequestering soil carbon and reducing greenhouse gas emissions, and finally increases crop production in a changing climate.

10.3 Philosophy and basic principles of conservation agriculture

Conservation agriculture facilitates good agronomy, such as timely operations, and improves overall land husbandry for rainfed and irrigated production. Complemented by other known good practices, including the use of quality seeds, and integrated pest, nutrient, weed and water management, etc., conservation agriculture is a base for sustainable agricultural production intensification. Conservation agriculture along with best crop management practices will lead to improvement in input use efficiency and greater sustainability (Gathalo et al., 2013; Wright et al., 2007; Vanden Bygaart et al., 2003).

A common philosophy among sustainable agriculture practitioners is that a 'healthy' soil is a key component of sustainability, that is, a healthy soil will produce healthy crop plants that have optimum vigour and are less susceptible to pests. Crop management systems that impair soil quality often result in greater inputs of water, nutrients, pesticides, and/or energy for tillage to maintain yields.

Regular additions of organic matter or the use of cover crops can increase soil aggregate stability, soil tilth and diversity of soil microbial life. Reducing tillage and increasing crop diversity to include perennial grasses and legumes could be very effective in carbon and N sequestration leading to sustainable production (Gathala et al., 2013; Alkaisi et al., 2005; Riedell et al., 2009). It reverses the land degradation, protects the environment and meets the challenges of climate change as a result the crop yield increases along with the profitability.

Conservation agriculture uses three interrelated principles such as (i) minimum mechanical soil disturbance, (ii) permanent soil cover with crop residues and mulches, and (iii) diversified crop rotation and cover cropping. The major functions are;

Prevention of soil degradation and erosion: To reduce soil disturbance, farmers practice zero-tillage farming, which allows direct planting without ploughing or preparing the soil.

Production of abundant above- and below-ground biomass to protect the soil: The soil receives physical protection from the weather (impact of raindrops, force of the wind and heat from solar radiation). The soil erosion is reduced and nutrient use efficiency is improved resulting better crop yield. Due to presence of soil organic matter, the soil microbes get a better environment and contribute to soil aggregate formation and stability.

Balancing of the C/N ratio during crop rotation: Crop rotation between cereals (high in carbon) and legumes (high in nitrogen) improves the soil properties, reduces pest load and enhances the nitrogen level. A superficial layer of organic matter on the soil improves the capture and use of rainfall through increased water absorption and infiltration and decreased evaporation from the soil surface.

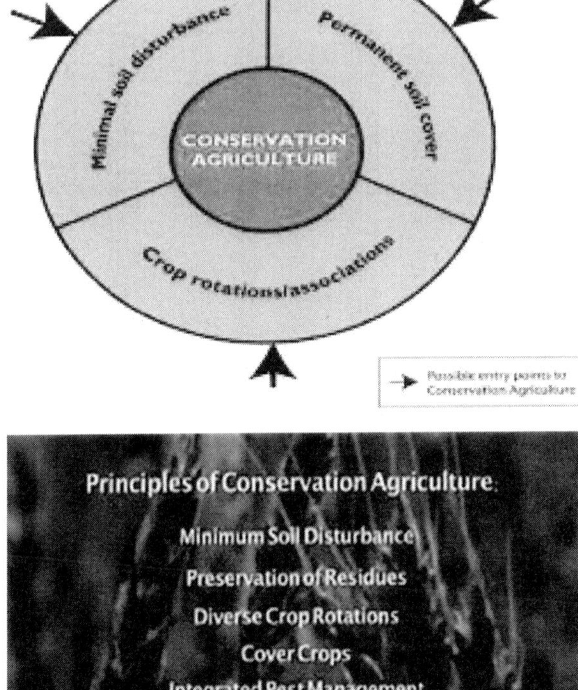

Fig. 10.2. Principles of conservation agriculture

10.4 Conservation agriculture practices

Conservation Agriculture can be adapted for various agro-ecological zones and socio-economic contexts and for many farming systems FAO). The broad principles and practices of CA are outlined below.

Principles	Practices
Conservation tillage or minimum tillage	Zero tillage/minimum tillage
Crop rotation	Inter cropping or crop rotation, improved seeds
Soil cover	Stubble and crop residues, management of natural flora, cover cropping

10.4.1 Soil management in conservation agriculture

Conservation agriculture adopts the following good soil management practices to improve the soil physical, chemical and biological properties for sustainable crop yield.

- Establish a detailed knowledge of the nature, properties, distribution, and potential uses of soils of the farm.
- Avoid mechanical soil disturbance to the extent possible.
- Avoid soil compaction beyond the elasticity of the soil.
- Maintain or improve soil organic matter during rotations until reaching an equilibrium level.
- Maintain organic cover through crop residues and cover crops to minimize erosion loss by wind and/or water.
- Maintain balanced nutrient levels in soils.

- Avoid contamination with agrochemicals, organic and inorganic fertilizers and other contaminants by adapting quantities, application methods and timing to the agronomic and environmental requirements.
- Maintain a record of the annual use and inputs and outputs of each individual land-management unit.

10.4.2 Maintaining and managing a permanent soil cover

Farm planning and crop rotations design

The planning of farm activities in Conservation Agriculture systems requires skills, knowledge and moreover, practices. It requires a 2-3 year crop rotation and proper management of cover crops to ensure sufficient biomass production to fulfil the needs of human beings and livestock. It should ensure that soil water and nutrient resources are adequate for the crop and that the use of agrochemical inputs such as desiccants and herbicides can be minimised. Crop rotation plays a critical role in conservation agriculture. It breaks the pest and disease cycle resulting higher crop yields. It helps the farmer in solving problems like soil compaction, plant diseases, weeds and slow early growth (Tarkalson et al., 2006).

Choice of cover crops

Multi-purpose crops like *Crotalaria juncea* should be chosen to meet the requirement of biomass production and nitrogen fixation. This cover crop is used in orchards or in rotation with many crops. The design of crop rotations and the choice and management of cover crops must ensure that the biomass production is sufficient to satisfy all the needs (permanent soil cover, human and livestock feeding, fibres, etc.), that soil water and nutrient resources are adequate for the crop. It should be also considered to avoid agro-chemicals. Cover crops must be multi-purpose crops and must be carefully managed through appropriate technologies. After harvest, crop residues are left in the field, usually without chopping them. Cover crops may be managed by using desiccants or by mechanical means such as knife rollers-recommended for small farmers. Mechanical tools have been developed by manufacturers, but many of them can be adapted by farmers themselves.

Management of crop residues and cover crops

The stubbles and crop residues which are left in soil after harvest of a crop act as soil cover and mulch. Mechanical tools such as knife rollers can also be used without harming the soil.

10.4.3 Minimum soil disturbance – zero tillage and direct planting

It is necessary to correct soil limitations such as hard pan, levelling, surface crusting, compaction and amending problematic soils before introducing zero tillage or minimum tillage. Direct seeding or planting crops after soil cover can be done depending on the specific conditions (soil, climate, seeds and cover properties). Careful selection of crops and crop rotations is necessary in consideration of soil and plant characteristics. The user friendly mechanical tools and equipment can be prepared locally for seeding and planting.

10.4.4 Pest, weed and soil fertility management

In conservation agriculture systems, pest and disease control are based on Integrated Pest Management (IPM) technologies. Weed control can be achieved by mulch cover and cover crops and crop rotations. Mechanical weed control through rolling, cutting or breaking can also be applied. Biological weed control may be very effective through appropriate choice of cover crops. Any soil limitations, such as acidity, salinity or toxicity problems must be addressed before and during the implementation of Conservation Agriculture. Soil Organic Matter (SOM) and judicious choice of crops and cover crops play a key role in soil fertility management, and especially SOM provided by root decomposition.

10.5 Benefits of conservation agriculture

Zero-tillage or minimum tillage with residue cover saves irrigation water, gradually increases soil organic matter and suppresses weeds, as well as reduces costs of machinery, fuel and time associated with tilling. Leaving the soil undisturbed decreases water infiltration, holds soil moisture and helps to prevent topsoil erosion. Conservation agriculture requires a new way of thinking about agricultural production in order to understand how one could possibly attain higher yields with less labour, less water and fewer chemical inputs. In spite of these challenges, conservation agriculture is spreading to farmers throughout the world as its benefits become more widely recognized by farmers, researchers, scientists and extension officers alike. Specifically, conservation agriculture (CA) increases the productivity of land, labour and capital.

Land - Conservation agriculture improves soil structure and protects the soil against erosion and nutrient losses by maintaining a permanent soil cover and minimizing soil disturbance. The practices enhance soil organic matter levels and nutrient availability.

Labour - Because land under no-till is not cleared before planting and involves less weeding and pest problems following the establishment of permanent soil cover/crop rotations, farmers save labour cost.

Water - Conservation agriculture enhances infiltration and water holding capacity due to crop residues which act as mulches. Organic mulches also protect the soil surface from extreme temperatures and greatly reduce surface evaporation, which is particularly important in tropical and sub-tropical climates.

Nutrients - Biochemical decomposition of crop residues increases soil nutrient supplies and cycling with the help of soil microbes. Fixation of nitrogen by legumes increases soil nitrogen level.

Soil biota - Insect pests and other disease causing organisms are kept under check by the activities of beneficial organisms including predatory wasps, spiders, nematodes, springtails, mites and beneficial bacteria and fungi, among other species. Soil property is increased due to activities of earth worms.

In general, soil fertility is built up over time under conservation agriculture, and fewer fertilizer amendments are required to achieve optimal yields over time. Insect pests and other disease causing organisms are held in check by an abundant and diverse community of beneficial soil organisms, including predatory wasps, spiders, nematodes, springtails, mites etc. The economic benefits of no-till and other conservation agriculture technologies, more than any other factor, have led to widespread adoption among both large- and small-scale farmers throughout the world. The organic carbon stock, biological activity, above- and below-ground biodiversity and soil structure are all improved due to conservation agriculture. Soil degradation – in particular soil erosion and run-off – is greatly reduced which increases crop yields. Reduced soil and nutrient losses, in combination with more rapid pesticide breakdown and greater adsorption also result in improved water quality. Carbon dioxide (CO_2) emissions are lowered as a result of the reduced use of machinery and increased accumulation of organic carbon.

References

Alkaisi, M.M., Yin, X. and Licht, M.A. 2005. Soil carbon and nitrogen changes as influenced by tillage and cropping systems in some Iowa soils. Agriculture, Ecosystems, and Environment, 105(4): 635-647.

Baker and saxena (2002). No tillage seeding science and practice, 2nd Edn,CAB, Oxford

FAO (2008). Investing in sustainable crop intensification; the case for soil health; Report of the international technical workshop, Rome.

Friedrich T, Derpach R, Kassam, AH (2012). Global overview of spread of conservation agriculture, Field Action, Sc Rep. 6: 1-7.

Gathala, M.K., Kumar, V., Sharma, P.C., Saharawat, Y.S., Jat, H.S., Singh, M., Kumar, A., Jat, M.L., Humphreys, E., Sharma, D.K., Sharma, S., and Ladha, J.K. 2013. Optimizing intensive cereal based cropping systems addressing current and future divers of agricultural changes in the north western Indo-Gangetic Plains of India. Agriculture, Ecosystems and Environment, 177(1): 85- 97.

Hobs, RP, Sayre K, Gupta R. (2008) The role of conservation agriculture in sustainable agriculture.

Pute et al (2010), Assessing the effect of soil tillage on crop growth. Eur J of Agron. 33: 231-241.

Recosky, DC, Saxena, KC (2007). The benefits of no tillage.

Riedell, W.E., Pikul, J.L., Jaradat, A.A. and Schumacher, T.E. 2009. Crop rotation and nitrogen input effects on soil fertility, maize mineral nutrition, yield and seed composition. Agronomy Journal, 101(4): 870-879.

Tarkalson, DD, Hergert GW, Cassaman KG (2006): Long term effect of tillage on soil chemical properties and grain yields of dryland winter wheat- sorghum/corn-fallow rotation in the Great Plains, Aron. J.98: 26-33.

Vanden Bygaart, A.J., Gregorich, E.G. and Angers, D.A. 2003. Influence of agricultural management on soil organic carbon: A compendium and assessment of Canadian studies. Canadian Journal of Soil Science, 83: 363-380.

Wright, A.L., Dou, F. and Hons, F.M. 2007. Soil organic C and N distribution for wheat cropping systems after 20 years of conservation tillage in central Texas. Agriculture, Ecosystems and Environment, 121: 376-382.

11

Climate Resilient Agriculture for Sustainability

11.1 Climate change

Climate change is a long-lasting change in the statistical distribution of weather pattern triggered by greenhouse gas (GHG) emission. It is caused by natural factors like continental drift, volcanoes, earth's tilt, and ocean current or man-made factors like urbanization, industrialization, burning of fossil fuel, deforestation and faulty agricultural practices. Agriculture is the second highest source of GSGs emission (19.6% of total emissions) (FAOSTAT). Faulty agricultural factors are use of chemical fertilizers, low nutrient use-efficiency pesticides, enteric fermentation, transplanted rice cultivation etc. Global warming is a serious consequence of climate change along with weather changes. Climate change has posed a serious threat to sustainable agriculture as it can reduce agricultural income by 15-25 per cent. It is therefore imperative that rationale of climate-resilient agriculture (CRA) is valued and implemented more rigorously.

11.2 Effects of climate change on agriculture

Agriculture is greatly influenced by weather and interlinked. The change in temperature, rainfall pattern affect the plant growth and production directly. The indirect effects are land availability, irrigation, weed growth, pest and diseases outbreak etc. High temperature reduces quality and yield of crops, and also increases pest and weed proliferation. In general, the temperate regions appear to be less vulnerable to climate change than the tropical regions due to the fact that higher temperatures in temperate areas shift biological process rates toward optima, and beneficial effects are likely to ensure (Rosenzweig, C., et al.1992). It is found that increase of temperature by 1° C decrease in grain yield of C_3 plant like rice by 6% (Saseendran., 2000) and in wheat, soybean, mustard, groundnut, potato by 3 to 7 % (Dagar et al., 2012). Increase in CO_2 to 550 ppm increase the yield C_3 plants like rice, wheat, legumes, and oil seed by 10-20% (Venkateswarlu., 2014). In general, vegetative growth is positively co-related with elevated CO_2 level but the reproductive stage of the crop is more linked with an optimum temperature affecting economic yield. Change in precipitation pattern increases the probability of short-run crop failures and long-run production. Changes in precipitation pattern increase the pest infestation.

11.3 Resilience to climate change

Resilience is the ability of a system and its component to anticipate, absorb, accommodate or recover from the effect of hazardous event in a timely and efficient manner (IPCC, 2012). Adverse effects of global warming include reduced crop quantity and quality due to the reduced growth period following high levels of temperature rise, reduced sugar content, and reduced storage stability in fruits, increase of weeds, blights, and harmful insects in agricultural crops. Climate resilient agriculture increases the capacity of the system to bounce back to the previous situation. Climate resilient agriculture is present in nature, but the pace is slower than the rapid climate change. Climate-resilient agriculture (CRA) is an approach that includes sustainable use of existing natural resources through crop and livestock production systems to achieve long-term higher productivity and farm incomes under climate variability. Climate resilience in agriculture can be achieved by adaptation or mitigation measures.

11.3.1 Improved techniques for adaptation to climate change

Weather based agro-advisories: The agro advisory based on weather parameters such as rainfall, temperature, relative humidity and wind speed etc. can be displayed in public domain to improve the weather literacy of the farmers. Mobile phones are being used to give the personal message to the farmer for short coming weather condition and it is now ever-increasing appeal to rural users. Weather based agro advisory helps the farmer to take up all possible measures in time.

Smart crop and variety selection: Selection of a climate-smart crop and variety to withstand weather calamities like heat/cold wave, flood, cyclone, frost, hail storm etc. is ideal to face climate change. In rice-wheat cropping system introduction of a short duration summer legume like moong bean as catch crop after wheat harvesting maintain the soil quality as well as it adds some organic matter and reduce the N_2O emission from the field.

Efficient climate-based cropping system: Mixed cropping, intercropping, relay cropping reduces the climatic vulnerabilities. Farmers can get at least one crop under adverse weather. Inclusion of a legume in cropping system adds sustainability to this system through soil cover and addition of biological nitrogen.

Water harvesting: An increase in temperature will increased demand for water for evapotranspiration by crops and natural vegetation and will lead to more rapid depletion of soil moisture. According to one projection, a rise in 1°C will increase the crop water demand by 2%. Increased precipitation induces run off, whereas increase in temperature may enhance the evapotranspiration demand. Hence, rain water harvesting and conservation are required in dryland areas which can be used for supplementary irrigation and lifesaving irrigation under moisture stressed period.

Water management: Water-smart technologies like a furrow-irrigated raised bed, micro-irrigation, rainwater harvesting structure, cover-crop method, greenhouse, laser land levelling; reuse wastewater, deficit irrigation and drainage management can support farmers to decrease the effect of variations of climate. Cultivating less water requiring varieties, adjusting the planting dates, irrigation scheduling, and adopting zero-tillage which may help farmers to reach satisfactory crop yields, even in deficit rainfall and warmer years.

Balanced fertilization: Balanced application of NPK based on soil test provides optimum plant growth with highly efficient nutrient use which can reduce adverse effect on environment. Balanced fertilization reduces the N_2O-N emission by controlled use of nitrogen fertilizer.

Contingency planning: Contingency crop planning for reverent rainfall refers to planning for alternate, crop and cultivator to suite the resource endowments of rainfall and soil in a given location (Reddy., 2019). In rain fed areas as a general rule, early showing of crop with the onset of monsoon is the best-bet practises for obtaining maximum yield.

11.3.2 Improved techniques to mitigate climate change

The adverse impact of climate change can be mitigated by reduction of food losses and waste, improved crop management practices, recarbonization of soil, no-till farming, site specific nutrient management, integrated farming system etc.

Reduction of food losses and waste: Along with food production it is necessary to reduce food wastage. Food wastage can be reduced at community level and household level. In the home, one of the best ways to reduce food waste is to plan meals ahead, rotate time-sensitive foods in the fridge and cupboards and freeze surplus garden vegetables.

Improved crop management practices: Intermittent irrigation reduces the total carbon flux in rice fields. To reduce CH_4 production from rice field, direct seeded rice (DSR), alternate wetting and drying (AWD) techniques are found effective.

Recarbonization of soils: Soil organic carbon management through increasing soil carbon storage can increase infiltration, increase fertility and nutrient cycling, decrease wind and water erosion, minimize compaction, enhance water quality, and generally enhance environmental quality. Practices like residue management, eliminating fallow period by permanent plant cover in soil, diversified crop rotation with legume, agroforestry etc. are recommended for this. N source from the legume play an important role to control the carbon sequestration. Agroforestry is also a sound practice for recarbonization.

No-Till system: No-till system can the physical properties of soil and the greenhouse gas (GHG) balance. Farmers can save labour and fuel costs, reduce soil erosion and preserve precious nutrients by adopting this practice. No-till also increases the accumulation of soil organic carbon, thereby resulting in sequestration of atmospheric carbon dioxide.

Site specific nutrient management: Site Specific Nutrient Management (SSNM) is an approach of supplying plants with nutrients to optimally match their inherent spatial and temporal needs for supplemental nutrients by using of SSNM through right amount, right source, right rate of application, right time, and right method. Devises like chlorophyll meter (SPAD meter), leaf colour chart (LCC) and green seeker are used for site specific nutrient management. Efficient N management can help in adaptation and mitigation while reducing other environmental threats such as eutrophication, acidification, air quality and human health.

Integrated farming system (IFS): Integrated farming system is more climate resilient than mono-cropping. IFS is often less risky, because it manages the farm more efficiently without dependence on external inputs., thus reduce the dependence of output. Intermittent use of farm produces proper recycling on by-products, crop residue, weed, an all-other farm waste combined with conservation of farm resource have been found to reduce chemical load in the form of inorganic fertiliser by 36% (Gangwar et al., 2014).

References

Dagar, J.C., Singh, A.K., Rajbir-Singh and Arunachalam, A. Climate change vis-a-vis Indian agriculture. Annals of Agricultural Research New Series, vol. 33,no. 4, pp 189-203. 2012.

Food and Agriculture Organization of the United Nations Statistics Division (FAOSTAT). India, Emissions – Agriculture total, viewed on September 18, 2018.

Gangwar, B. and Singh, J.P. Integrated Farming Systems ResearchConcepts and Status. (In) Research in Fanning Systems, Gangwar, B., Singh, J.P., Prusty, A.K. and Prasad, K. (Eds). Today and Tomorrow's Printers and Publisher, New Delhi. pp. 1-34. 2014.

.IPCC. 2012. Managing the risks of extreme events and disasters to advance climate change adaptation. A special report of working groups i and ii of the inter-governmental panel on climate change, pp-582. 2012.

Prasad, YG., Maheshwari, M., Dixit, S., Srinivasarao, Ch., Sikka, AK., Venkateswarlu, B., Sudhakar, N., Prabhu Kumar, S., Singh, AK., Gogoi, AK., Singh, AK., Singh, YVand Mishra, A. Smart Practices and Technologies for Climate Resilient Agriculture. Central Research Institute for Dryland Agriculture (ICAR), Hyderabad. 2014.

Reddy Sr., Principles of Agronomy, pp. 244-300. 2019.

Rosenzweig, C., and D. Liverman. Predicted effects of climate change on agriculture: A comparison of temperate and tropical regions. In Global climate change: Implications, challenges, and mitigation measures, ed. S. K. Majumdar, PA: The Pennsylvania Academy of Sciences, pp. 342-361. 1992.

Saseendran, A.S.K., Singh, K.K., Rathore., LS Singh., S.V. and Sinha, S.K. Effects of climate change on rice production in the tropical humid climate of Kerala, India. Climate Change, vol. 44, pp. 495—514. 2000

Shaun A. Marcott et al, A Reconstruction of Regional and Global Temperature for the Past 11,300 Years, Science vol. 339, pp. 1198. 2013.

Venkateswarlu, B et all. Climate resilient agronomy: an overview, Indian society of agronomy, new Delhi. pp 1-11. .2016.

12

Summary of Sustainable Agriculture Practices

12.1 Rationale of sustainable agriculture

Agriculture is a fundamental human activity that depends intrinsically on natural processes, including soil fertility, water recycling, and pollination, and both nature and agriculture are increasingly suffering the negative impacts of climate change (EEA, 2019). At the same time, unsustainable agriculture also poses a major threat to biodiversity (Secretariat of the Convention on Biological Diversity, 2014), negatively affects the state of our soil and water, and is an important contributor to climate change (Jia et al., 2019). In the European Union (EU), agriculture is the activity most frequently cited as negatively impacting the state of nature (EEA, 2015).

With the climate change impacting agriculture the concept of " sustainable agriculture" is emerging as a necessity. This is also important in the context of the nature-based solutions as they have been defined: "actions to protect, sustainably manage and restore natural or modified ecosystems, which address societal challenges (e.g. climate change, food and water security or natural disasters) effectively and adaptively, while simultaneously providing human well-being and biodiversity benefits" (Cohen Shacham et al., 2016).

Sustainable agriculture should be in agreement with the definition of sustainable development. As outlined in the UN's 1987 Brundtland report (WCED, 1987), sustainable agriculture should be able to meet the current needs of society without compromising the ability of future generations to meet their own needs. It should take into account environmental, social, and economic sustainability, which constitute the three central pillars of sustainable development. FAO defines sustainable agriculture as the "management and conservation of the natural resource base, and the orientation of technological and institutional change in such a manner as to ensure the attainment and continued satisfaction of human needs for present and future generations. FAO has proposed five principles (FAO, 2014) for sustainable agriculture that capture all three pillars, namely: 1) improving efficiency in the use of resources, 2) conserving, protecting and enhancing natural ecosystems, 3) protecting and improving rural livelihoods and social wellbeing, 4) enhancing the resilience of people, communities, and ecosystems, and 5) promoting good governance of both natural and human systems. In another attempt to define sustainable agriculture, the United Kingdom's Royal Society (2009) enumerates four principles for agricultural sustainability: persistence, resilience, autarchy and benevolence. It states that any approach is unsustainable if it depends on non-renewable inputs, cannot consistently and predictably delivers desired outputs, can only do this by requiring the cultivation of more land, and/or causes adverse and irreversible environmental impacts.

12.2 Approaches to SA

There are various approaches recommended for sustainable agriculture.

12.2.1 Agro-ecology

Agro-ecology is designed and managed through the implementation of a structural and functional diversification of the biological components of production systems, such as intercropping, polycultures, crop-livestock integration, agroforestry, and multispecies livestock keeping (FAO, 2019). Important practices adopted in agro-ecology are: conservation tillage, inter cropping/ mixed cropping, crop rotation, cover cropping, crop-livestock integration, integrated nutrient management, biological control of pests, water harvesting, agroforestry, use of renewable energy and holistic land management.

12.2.2 Nature-inclusive agriculture

Nature-inclusive agriculture also considers the farming system as an agro-ecosystem, focusing on its sustainability. Nature-inclusive agriculture aims to minimise the negative effects of agriculture on nature and to maximise the positive effects of nature on agriculture. The practices adopted in this system are; low cattle densities, less inorganic fertiliser use, reduced tillage and enlarging diversity of landscape.

12.2.3 Permaculture

In Mollison's words, "Permaculture is a philosophy of working with, rather than against nature; of protracted and thoughtful observation rather than protracted and thoughtless labour; and of looking at plants and animals in all their functions, rather than treating any area as a single product system" (Mollison, 1991). The practices adopted in permaculture are: rainwater harvesting, composting, making use of ecosystem services (nitrogen fixer, pollinator), building soil health, no-tilling, use of no chemicals, agroforestry, and using animals for multiple functions.

12.2.4 Bio-dynamic agriculture

Biodynamic agriculture may be defined as an ecological farming system that views the farm as a self-contained and self-sustaining organism; farmers strictly avoid all synthetic pesticides and fertilisers, instead using living solutions for pest control and fertility, and set aside a minimum of 10% of the total farm's area for biodiversity (Demeter, 2012). The usual practices are; treating the farm as a living organism, conservation of bio-diversity, preparation of special bio-dynamic preparations and composts, and integration of crop-livestock for nutrient management. There are six preparations made from yarrow, chamomile, stinging nettle, oak bark, dandelion, and valerian, the application of which fosters the growth of beneficial bacteria and fungi and enriches the development of the compost, stabilising nitrogen and other nutrients, multiplying microbial diversity, and helping sequester carbon. Preparation of horn manure (made from cow manure buried inside a cow horn during the winter months) enhances the life of the soil, horn silica (prepared from ground quartz crystals buried in a cow horn over the summer months) increases plant immunity, strengthens photosynthesis, and enhances ripening, and horsetail tea helps prevent fungal diseases.

12.2.5 Organic farming

IFOAM's definition for organic farming outlines it as "a production system that sustains the health of soils, ecosystems and people. It relies on ecological processes, biodiversity, and cycles adapted to local conditions, rather than the use of inputs with adverse effects. Organic agriculture combines tradition, innovation and science to benefit the shared environment and promotes fair relationships and a good quality of life for all involved." (IFOAM, 2005). The important practices in organic farming area : crop rotation, no use of chemical Fertilisers, herbicides and pesticides, use of resistant crop varieties and breeds, prohibition of GMO varieties, prohibition of hormones and antibiotics, and use of organic matter for soil health management.

12.2.6 Conservation Agriculture

Through the conservation of soil, the approach contributes to enhancing biodiversity both within and above the soil, capturing more carbon, enabling a higher efficiency of water and nutrient use, and ultimately resulting in improved and sustained food production (FAO). The practiced adopted in conservation agriculture are: minimum soil disturbance with reduced or no-tillage, maintenance of soil cover by crop residues, mulches and cover cropping, and managing crop rotation with crop diversity.

12.2.7 Regenerative agriculture

It consists of practices that regenerate soil, reducing but not necessarily eliminating synthetic pesticides and fertilisers, and going beyond the reduction of negative effects towards ensuring that agriculture has a positive effect on the environment. The practices adopted in this system are: no-till or minimum tillage, managing soil fertility by cover cropping, crop rotation, composting and animal manures, diversifying cropping system with inter cropping, mixed cropping, multi-species cover cropping, agro-forestry, silvopastoral system, grazing management and increasing soil carbon and soil carbon sequestration.

12.2.8 Carbon farming

Based largely on the principles of conservation and regenerative agriculture, carbon farming "involves implementing practices that are known to improve the rate at which CO_2 is removed from the atmosphere and converted to plant material and/or soil organic matte" (Carbon Cycle Institute). Its key goal is to store carbon beneficially in soils as well as vegetation, stemming from a sense of opportunity and responsibility with respect to the agricultural sector‘s role in climate change. The usual practices area: residue management, no-till farming with direct seeding, multi-storey cropping, alley cropping, mulching, tree planting and nutrient management by mulching and composting. Vegetative barrier, filter strips and hedge rows are also recommended.

12.2.9 Climate smart agriculture

According to the FAO, climate smart agriculture aims to enhance the capacity of the agricultural systems to support food security, incorporating the need for adaptation and the potential for mitigation into sustainable agriculture development strategies. The climate smart agricultural practices are: integrated crop-livestock production system and agro-forestry, selection of adapted crops and varieties including animal breeds to climate change, efficient farming practices to reduce emission of GHG, animal and grazing management, sustainable forest management, sustainable land and water management, use of agro-advisory services, renewable energy management, reduction of fossil energy consumption, site specific nutrient management and precision agriculture etc.

12.2.10 High nature value farming

According to Oppermann et al. (2012), the HNV farming approach stipulates that, being at the lowest end of the farming intensity spectrum, not only the farmland margins but also the productive land itself supports a range of wildlife species absent from intensively farmed land. The practices adopted in this system are: grazing management, organic farming, mixed cropping, hay meadows, conservation of nature, fallowing etc.

12.2.11 Low external input agriculture

Low external input agriculture (LEIA) is described as an approach referring to a set of agronomic practices that aim to reduce the use of inputs from outside the production system. The use of green and animal manures, replacing fertilisers, limiting use of herbicides and pesticides, biological pest management, crop rotation, inter cropping, cover cropping, diversified cropping and introduction of legumes for nitrogen fixation are usually adopted in this system.

12.2.12 Circular agriculture

Circular agriculture takes a whole system approach, involving the integration of crops and livestock and making the best possible use of resources, including side streams (de Boer & van Ittersum, 2018), indicating a shift from production-efficiency to resource-efficiency. The practices adopted here are precision agriculture, breeding for nitrogen fixing varieties, residue management, fodder cultivation, organic recycling for production of manure, inter cropping, sustainable biomass production, and measures for mitigating GHG emission.

12.2.13 Ecological intensification

Ecological intensification is an approach that aims to match or increase agricultural production as compared to conventional farming methods, while minimising negative impacts on the environment and on agricultural productivity, by integrating the management of ecosystem services delivered by biodiversity into production systems (Bommarco et al., 2013). The practices adopted in this system are; mixed cropping, diversified crop rotation, use of cover crops, direct-seeding, and mulch-based cropping , conservation tillage, minimising soil compaction and soil detoxification, integrated pest management, improved nutrient management, fertigation, and biodiversity preservation.

12.2.14 Sustainable intensification

The approach may involve the intensification of different types of agricultural inputs (e.g. of knowledge, biotechnologies, labour, machinery) and apply these to different forms of agriculture (e.g. livestock or arable; agro-ecological or conventional) (FCRN Food source). The practices adopted in this system are: precision farming, using robotics, artificial intelligence and big data, integrated crop-livestock farming, vertical farming, hydroponics, aeroponics, biotechnology, greenhouse technology etc.

References

Bommarco, R., Kleijn, D., Potts, S. G. (2013). Ecological intensification: harnessing ecosystem services for food security. Trends in Ecology & Evolution. 28:4. 230- 238 pp., ISSN 0169-5347. Available at: https://doi. org/10.1016/j.tree.2012.10.012.

Cohen-Shacham, E., Walters, G., Janzen, C. and Maginnis, S. (eds.) (2016). Nature-based Solutions to address global societal challenges. Gland, Switzerland: IUCN. xiii + 97pp. DOI: https://doi. org/10.2305/IUCN.CH.2016.13.en World Commission on Environment and Development (WCED) (1987). Our Common Future. Oxford University Press, 383 pages

de Boer, I.J.M., van Ittersum, M.K. (2018). Circularity in agricultural production. Scientific basis for the Mansholt lecture 2018. Wageningen University and Research. Available at: https://edepot.wur.nl/470625.

Demeter. Biodynamic preparations. [website]. Available at: https://www.demeter.net/ what-is-demeter/ biodynamic-preparations.

EEA (2019). Climate change threatens future of farming in Europe. [online news]. Available at: https:// www.eea.europa.eu/highlights/climate-changethreatens-future-of?utm medium= email&utm_campaign=Climate%20change%20hreatens%20future %20of% 20farming %20in%20 Europe&utm_content=Climate%20change%20 threatens%20future%20of%20farming%20in%20Europe+CID_5ff0d 97bb588d3762bbc8d28e8 7df206&utm_source =EEA%20Newsletter&utm_ term=Read%20more#tab-related-publications.

European Commission's DG Agriculture and Rural Development (DG AGRI), International Union for Conservation of Nature (IUCN), (10th December 2018, 9th-10th October 2019). Round tables on the green architecture of the CAP. [event] Available at: https://ec.europa.eu/info/events/ round-tables-greenarchitecture-cap-2019-oct-09_en ; https://ec.europa. eu/info/events/round-tables-green-architecture-cap2018-nov-12_en.

European Environment Agency (EEA) (2015). State of nature in the EU: biodiversity still being eroded, but some local improvements observed. [online news]. Available at: http://www.eea.europa.eu/highlights/ state-of-nature-in-the.

FAO(2019). The State of the World's Biodiversity for Food and Agriculture, J. Bélanger & D. Pilling (eds.). FAO Commission on Genetic Resources for Food and Agriculture Assessments. Rome. 572 pp. Available at: http://www.fao.org/3/CA3129EN/ CA3129EN.pdf.

FAO, International Atomic Energy Agency (IAEA). Climate smart agriculture. Nuclear techniques. [website]. Available at: http://www.fao.org/climatesmart-agriculture/knowledge/practices/nucleartechniques/en/.

FAO. (2014). Building a common vision for sustainable food and agriculture. Principles and approaches. Rome.

FCRN Foodsource. Climate smart agriculture - glossary. [website]. Available at: https://foodsource.org.uk/ glossary/c.

Food and Agriculture Organization of the United Nations (FAO) (1988). Report of the FAO Council, 94th Session, 1988. Rome

Intergovernmental Science-Policy Platform on Biodiversity and Ecosystem Services (IPBES) (2019). Summary for policymakers of the global assessment report on biodiversity and ecosystem services of the Intergovernmental Science-Policy Platform on Biodiversity and International Federation of Organic Agriculture Movements (IFOAM) (2005). Definition of organic agriculture. [website]. Available at: https://www. ifoam.bio/fr/organic-landmarks/definition-organic agriculture.

Jia, G. et al. (2019). Land–climate interactions. In: Climate Change and Land: an IPCC special report on climate change, desertification, land degradation, sustainable land management, food security, and greenhouse gas fluxes in terrestrial ecosystems [P.R. Shukla, et al (eds.)]. In press. Available at: https://www.ipcc.ch/site/assets/uploads/ sites/4/2019/11/05_Chapter-2.pdf.

Mollison, B. (1991). Introduction to permaculture. Tasmania, Australia: Tagari. Available at: https:// www.academia.edu/5098938/Introduction_To_ Permaculture_-_Bill_Mollison.

Oppermann, R., Beaufoy, G., Jones, G. (eds.) (2012). High Nature Value Farming in Europe: 35 European countries –experiences and perspectives. c. Verlaggegionalkultur, Germany. 544pp. ISBN:978- 89735-657-3.

Secretariat of the Convention on Biological Diversity (2004) The Ecosystem Approach, (CBD Guidelines). Montreal. Secretariat of the Convention on Biological Diversity 50 p. Available at: https://www.cbd.int/doc/ publications/ea-text-en.pdf.

The Royal Society (2009). Reaping the Benefits: Science and the Sustainable Intensification of Global Agriculture. London. Available at: https://royalsociety.org/~/media/ Royal_Society_Content/policy/ publications/2009/4294967719.pdf.